Activities Manual

for

Programmable
Logic Controllers

Third Edition

Frank D. Petruzella

 **Higher Education**

Boston   Burr Ridge, IL   Dubuque, IA   Madison, WI   New York   San Francisco   St. Louis
Bangkok   Bogotá   Caracas   Kuala Lumpur   Lisbon   London   Madrid   Mexico City
Milan   Montreal   New Delhi   Santiago   Seoul   Singapore   Sydney   Taipei   Toronto

The McGraw·Hill Companies

Activites Manual for
PROGRAMMABLE LOGIC CONTROLLERS, THIRD EDITION
Frank D. Petruzella

Published by McGraw-Hill Higher Education, an imprint of The McGraw-Hill Companies,
Inc., 1221 Avenue of the Americas, New York, NY 10020. Copyright © 2005, 1998, 1989
by The McGraw-Hill Companies, Inc. All rights reserved.

 This book is printed on recycled, acid-free paper containing 10% postconsumer waste.
RECYCLED

7 8 9 0 QPD QPD 0 9 8

ISBN 978–0–07–829855–4
MHID 0–07–829855–5

www.mhhe.com

CONTENTS

24-1. The layout of Fig. 1-3 is that of a(n): **24-1.** _____
   a) relay schematic.   c) input module wiring diagram.
   b) ladder logic program.   d) output module wiring diagram.

24-2. In the circuit of Fig. 1-3, to energize the starter coil: **24-2.** _____
   a) the pressure switch, temperature switch, and manual
      pushbutton must all be closed.
   b) the pressure switch, temperature switch, or manual
      pushbutton must be closed.
   c) the pressure switch and temperature switch or the manual
      pushbutton must be closed.
   d) the pressure and temperature switches or the manual
      pushbutton and temperature switch must be closed.

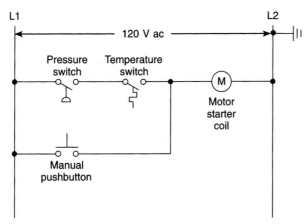

Fig. 1-3

25-1. The layout of Fig. 1-4 is that of a(n): **25-1.** _____
   a) relay schematic.   c) input module wiring diagram.
   b) ladder logic program.   d) output module wiring diagram.

25-2. In the diagram of Fig. 1-4, whenever there is a conducting path from **25-2.** _____
   left to right across the rung:
   a) I/1, I/2, and I/3 will all be at logic 1.
   b) I/1, I/2, and I/3 will all be at logic 0.
   c) I/1 and I/2 or I/3 will be at logic 1.
   d) I/1 and I/2 or I/3 will be at logic 0.

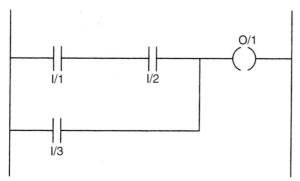

Fig. 1-4

26. Normally, on larger PLC systems, the purpose of the PLC power supply module is to power the:
a) field devices.
b) input devices.
c) output devices.
d) PLC modules.

26. _____

27. Which module of the PLC is responsible for performing logical operations?
a) Processor
b) Input
c) Output
d) Power supply

27. _____

28. Which module of the PLC connects to field devices such as pilot lights, motor starters, and solenoids?
a) Input
b) Output
c) Power supply
d) Memory

28. _____

29. _____ I/Os are typical of small PLCs that come in one package with no separate removable units.
a) Fixed
b) Modular
c) Digital
d) Analog

29. _____

30. PLC software that runs on a personal computer can be used to:
a) write a PLC program.
b) document a PLC program.
c) monitor and control the process.
d) all of the above.

30. _____

31. A control management PLC application normally requires a:
a) micro-size PLC.
b) small-size PLC.
c) medium-size PLC.
d) large-size PLC.

31. _____

32. Which of the following is *not* a factor effecting the memory size needed for a particular PLC installation?
a) Voltage rating of field devices
b) Number of I/O points
c) Size of control program
d) Supervisory functions required

32. _____

TEST 1•2

Place the answers to the following questions in the answer column at the right.

Answer

1. Programmable logic controllers were originally designed to perform logic functions previously accomplished by _____.

1. _____

2. The design of a PLC is similar to that of an electromechanical relay. (True or False)

2. _____

3. Match each of the following features of programmable logic controllers with the benefit that it offers.

3a. _____

3b. _____

FEATURES
1) solid-state components
2) small size
3) programmed circuits
4) modular design
5) "electrician-friendly"

3c. _____

3d. _____

3e. _____

BENEFITS
a) minimal space requirements
b) simple programming language
c) high reliability
d) expandability
e) easily modified

4. The CPU unit of a programmable logic controller contains a(n) _____.

4. _____

5. In a PLC system, there is physical connection between field input devices and output devices. (True or False)

5. _____

6. A(n) _____ device is required to enter the control program into the memory of the processor.

6. _____

7. Identify the following electrical components by specifying whether they are (1) input field devices or (2) output field devices.
a) pushbutton   c) pilot lamp
b) solenoid   d) selector switch

7a. _____

7b. _____

7c. _____

7d. _____

8. In a typical PLC ladder logic program the symbols represent the (a) _____ and the numbers represent the (b) _____.

8a. _____

8b. _____

9. The scan time is the time required for one complete execution of the user program. (True or False)

9. _____

10. The input/output system forms the interfaces through which field devices are connected to the controller. (True or False)

10. _____

11. Troubleshooting is simplified by the design of most PLCs because they include _____ indicators.

11. _____

12. _____ is the process of reading inputs, executing the program, and setting outputs on a continuous basis.

12. _____

13. The PLC is a much more complex computing machine than a computer. (True or False)

13. _____

14. Categorization of PLC size is primarily based on its I/O points. (True or False)

14. _____

15. Unlike the computer, a PLC can be programmed using relay ladder logic. (True or False)

15. _____

16. Quick disconnects in the I/O interfaces allow modules to be easily connected and replaced. (True or False)

16. _____

17. The _____ feature provides the single greatest benefit of PLCs over hardwired control.

17. _____

18. To operate the program, the controller is placed in the _____ mode.

18. _____

19. If there is no conducting path from left to right on the program rung, the output coil memory is set to logic _____.

19. _____

20. Changes to hardwired relay control systems usually require some _____ of the system.

20. _____

21. When programming the controller, instructions are entered _____ into the processor memory.

21. _____

22. The programming device must be connected to the controller to run the program. (True or False)

22. _____

23. Input and output devices are also referred to as _____ inputs and outputs.

23. _____

24. PLC systems usually require as much space in an enclosure as equivalent relay systems. (True or False)

24. _____

25. The term *central processing unit is* often used interchangeably with _____.

25. _____

26. What is the name of the most common programming language used in PLCs?

26. _____

27. Which PLC module connects to field devices such as pushbutton switches?

27. _____

28. The ladder logic program is stored in the processor module's _____.

28. _____

29. A programmable controller is basically a computer designed for use in machine control applications. (True or False)

29. _____

30. Problem solving with PLCs is a major advantage over relay-type control systems. (True or False)

30. _____

31. The programmable controller operates in real time. (True or False)

31. _____

32. When a module is slid into a PLC rack, it makes electrical connection with the _____.

32. _____

33. One disadvantage of modular I/O is its lack of flexibility. (True or False)

33. _____

34. On large PLC systems, the power supply module does not normally supply power to the field devices. (True or False)

34. _____

35. Removing the programming unit from the PLC will not affect the operation of the user program. (True or False)

35. _____

36. The PLC requires a CD drive. (True or False)

36. _____

37. Software written and run on a personal computer can be used to write a PLC program. (True or False)

37. _____

38. A _____ PLC application involves one PLC controlling one process.

38. _____

39. The _____ for a particular PLC lists all instructions along with a short description of their operation.

39. _____

PROGRAMMING ASSIGNMENTS FOR CHAPTER 1

1-1. On a separate sheet of paper, draw an I/O wiring diagram and ladder logic program for the relay schematic shown in Fig. 1-5. (Use the address numbers of the PLC simulator you will be working with.)

1-2. Have your instructor give you detailed instructions for placing the controller in the program mode and entering the ladder logic program into the memory.

1-3. Have your instructor give you detailed instructions for placing the controller in the run mode and running the program.

1-4. Have the operation checked by your instructor.

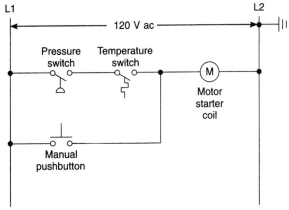

Fig. 1-5

2-1. On a separate sheet of paper, draw a ladder logic diagram for the modified relay ladder schematic shown in Fig. 1-6.

2-2. Have your instructor give you detailed instructions for clearing the original program from the memory of the PLC and entering the modified program into memory.

2-3. Have the operation checked by your instructor.

3-1. On a separate sheet of paper, draw a modified ladder logic program of the original circuit altered so that the manual pushbutton, pressure switch, and temperature switch all must be closed to permit operation of the motor.

3-2. Enter the program into the PLC and prove its operation.

4-1. On a separate sheet of paper, draw a modified ladder logic program of the original circuit altered so that the motor will operate when either the manual pushbutton, pressure switch, or temperature switch is closed.

4-2. Enter the program into the PLC and prove its operation.

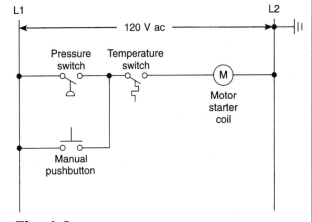

Fig. 1-6

2 PLC HARDWARE COMPONENTS

TEST 2•1

Choose the letter that best completes the statement.

Answer

1. One function of a PLC input interface module is to:
   a) accept signals from the process field devices and convert them into signals that can be used by the processor.
   b) convert signals from the processing unit into values that can be used to control the machine or process.
   c) input signals from the programming device and convert them into signals that can be used by the CPU.
   d) interpret and execute the user program that controls the machine or process.

1. _____

2. The location of a specific input or output field device is identified by the processor by means of its:
   a) voltage rating.   c) wattage rating.
   b) current rating.   d) address.

2. _____

3. A discrete output interface module is designed to provide:
   a) output voltages in the 5 V dc range.
   b) varying ac or dc voltages depending on the type of module selected.
   c) simple ON/OFF switching control.
   d) binary-coded outputs.

3. _____

4. The following statement that does *not* apply to the optical isolator circuit used in I/O modules is that it:
   a) separates high voltage and low voltage circuits.
   b) rectifies ac signals.
   c) prevents damage caused by line voltage transients.
   d) reduces the effect of electrical noise.

4. _____

5. Individual outputs of a typical ac output interface module usually have a maximum current rating of about:
   a) 1 A.   c) 500 mA.
   b) 50 A.   d) 15 μA.

5. _____

6. Which of the following input field devices would most likely be used with an analog interface input module?
   a) Pushbutton   c) Selector switch
   b) Limit switch   d) Photocell

6. _____

7. The "ON state input voltage range" specification refers to: **7.** _____
    a) the type of voltage device that will be accepted by the input.
    b) range of leakage voltage present at the input in its ON state.
    c) minimum and maximum output operating voltages.
    d) voltage at which the input signal is recognized as being ON.

8. Which of the following is *not* specified for discrete interface input   **8.** _____
    modules?
    a) Maximum surge current   c) Input delay
    b) Nominal voltage   d) Electrical isolation

9. The term *central processing unit* (CPU) is often used interchangeably   **9.** _____
    with:
    a) *processor.*   c) *programming device.*
    b) *power supply.*   d) *user program.*

10. Volatile memory elements can be classified as those that: **10.** _____
    a) do not retain stored information when the power is removed.
    b) retain stored information when the power is removed.
    c) do not require a battery backup.
    d) both *b* and *c.*

11. _____ is designed to permanently store a program that cannot be   **11.** _____
    altered under any circumstances.
    a) RAM   c) EPROM
    b) PROM   d) EEPROM

12. Which of the following memory types is often referred to as   **12.** _____
    read/write memory?
    a) RAM   c) EPROM
    b) PROM   d) EEPROM

13. EPROM is: **13.** _____
    a) electrically erasable memory.
    b) electrically programmable RAM.
    c) erased by exposure to ultraviolet light.
    d) not programmable.

14. The most common form of memory used to store, back up, or transfer   **14.** _____
    PLC programs is:
    a) RAM   c) EPROM
    b) PROM   d EEPROM

15. Which of the following is *not* a function of a PLC programming device?   **15.** _____
    a) To enter the user program
    b) To change the user program
    c) To execute the user program
    d) To monitor the user program

16. Status indicators are provided on each output of an output module to indicate that the:
    a) load has been operated.
    b) input associated with the output is active.
    c) module fuse has blown.
    d) output is active.

16. _____

17. The I/O system provides an interface between:
    a) input modules and output modules.
    b) the CPU and field equipment.
    c) the CPU and I/O rack.
    d) the I/O rack and I/O modules.

17. _____

18. The PLC chassis comes in different sizes according to the:
    a) size of the program.   c) number of slots they contain.
    b) type of I/O modules used.   d) all of the above.

18. _____

19-1. Section No. 1 of the input module block diagram in Fig 2-1 is the:
    a) power section.   c) logic section.
    b) high voltage section.   d) both _a_ and _b_.

19-1. _____

19-2. The output from section No. 2 in the input module block diagram of Fig. 2-1 is:
    a) high voltage ac.   c) high voltage dc.
    b) low voltage ac.   d) low voltage dc.

19-2. _____

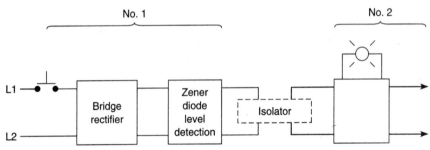

Fig. 2-1

20-1. The simplified schematic diagram of Fig. 2-2 is that of a(n):
    a) discrete output module.   c) discrete input module.
    b) analog output module.   d) analog input module.

20-1. _____

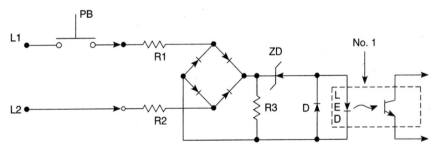

Fig. 2-2

20-2. The purpose of R1 and R2 in the schematic of Fig. 2-2 is to:
   a) change ac to dc.
   b) change dc to ac.
   c) lower the applied voltage.
   d) operate the status indicator light.

20-2. _____

20-3. The purpose of ZD in the schematic of Fig. 2-2 is to:
   a) switch the output circuit into conduction.
   b) set the minimum level of voltage that can be detected.
   c) protect against voltage transients.
   d) smooth out the rectified dc voltage.

20-3. _____

20-4. Part No. 1 in the schematic of Fig. 2-2 is the:
   a) bridge rectifier unit.
   b) optical isolator unit.
   c) filter circuit unit.
   d) processing/memory unit.

20-4. _____

21. Figure 2-3 shows the wiring connection of an output module terminal board that has been wired *incorrectly.* To correct the error, you would remove the wire from:
   a) L1 to terminal-board L1 connection.
   b) L1 to terminal-board 2 connection.
   c) terminal-board 2 connection to lamp.
   d) L2 to lamp connection.

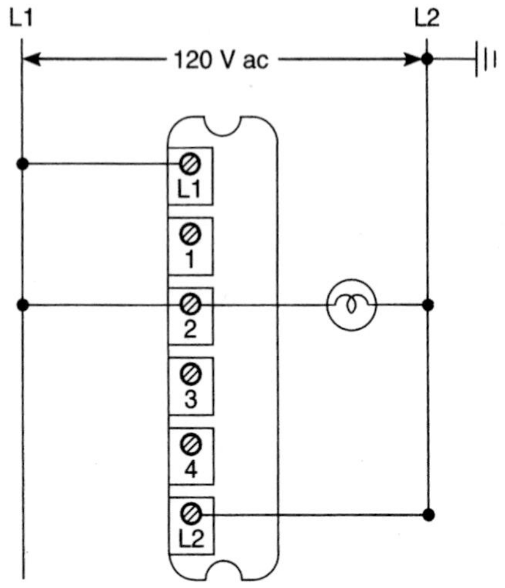

21. _____

Fig. 2-3

22-1. The simplified schematic diagram of Fig. 2-4 is that of a(n):
   a) discrete output module.   c) discrete input module.
   b) analog output module.   d) analog input module.

22-1. _____

22-2. The input signal in the schematic of Fig. 2-4 is obtained from:
   a) an input field device.
   b) an output field device.
   c) internal logic circuitry of the processor.
   d) either *a* or *b.*

22-2. _____

22-3. The purpose of the triac in the schematic of Fig. 2-4 is to: 22-3. _____
    a) switch current to the lamp ON and OFF.
    b) vary the current flow to the lamp in accordance with the input signal level.
    c) vary the voltage across the lamp in accordance with the input signal level.
    d) both *b* and *c*.

22-4. When the triac in the schematic of Fig. 2-4 is in the OFF state: 22-4. _____
    a) no current flows through the lamp.
    b) a small leakage current flows through the lamp.
    c) the rated surge current flows through the lamp.
    d) the rated nominal current flows through the lamp.

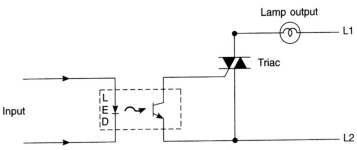

Fig. 2-4

23. The circuit of Fig. 2-5 is an example of how a PLC output is   23. _____
connected to:
    a) isolate the load from the controller.   c) vary the speed of a motor.
    b) control a high voltage circuit.   d) control a high current load.

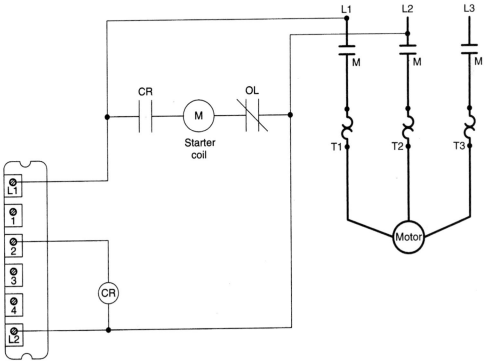

Fig. 2-5

24-1. The terminal connection diagram shown in Fig. 2-6 is that of a(n):

    a) discrete output module.
    b) analog output module.
    c) discrete input module.
    d) analog input module.

24-1. _____

24-2. Shielded cable is used in wiring the circuit of Fig. 2-6 to:

    a) reduce unwanted electrical noise signals.
    b) carry the higher current required.
    c) lower the resistance of the conductors.
    d) insulate the circuit from other cables.

24-2. _____

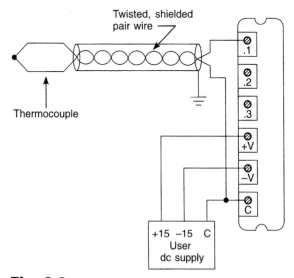

Fig. 2-6

25-1. In the block diagram of the central processing unit (CPU) shown in Fig. 2-7, blocks No. 1 and No. 2 represent the:

    a) processing and memory units.
    b) processing and power supply units.
    c) memory and programmer units.
    d) memory and power supply units.

25-1. _____

25-2. Block No. 3 in Fig. 2-7 represents the:

    a) processing unit.
    b) memory unit.
    c) power supply unit.
    d) programmer unit.

25-2. _____

25-3. Which of the following is the main function of the CPU unit?   25-3. _____
    a) Signal conditioning   c) Program and monitor system
    b) Signal isolation   d) Execute program

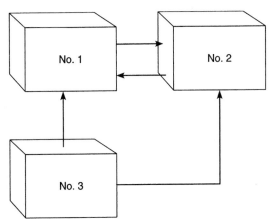

Fig. 2-7

26. A remote I/O rack: 26. _____
    a) is located away from the processor module.
    b) is used to minimize the amount of wiring required.
    c) requires the use of a communications network.
    d) all of the above.

27. High-density modules: 27. _____
    a) may have up to 32 inputs or outputs per module.
    b) require more space.
    c) can handle greater amounts of current per output.
    d) all of the above.

28. Discrete I/O modules can be classified as: 28. _____
    a) bit oriented.   c) processor oriented.
    b) word oriented.   d) power supply oriented.

29. Which of the following output module switching devices can be   29. _____
used to control dc devices?
    a) Transistor   c) Relay
    b) Triac   d) Both *a* and *c*

30. The design of _____ field devices typically requires that they be used in a specific sinking or sourcing circuit.
a) input
b) output
c) dc
d) ac

31. The _____ of an analog I/O module specifies how accurately an analog value can be represented digitally.
a) number of inputs and outputs per card
b) input impedance and capacitance
c) resolution
d) common mode rejection ratio

32. The processor section of the PLC is where the:
a) ladder logic program is stored.
b) input connections are made.
c) output connections are made.
d) sensors are located.

33. When placed in the _____ mode, the processor does not scan/execute the ladder program.
a) program
b) run
c) test
d) remote

34. Typically, a word stored in PLC memory consists of:
a) 8 bytes.
b) up to 24 letters.
c) 16 bits.
d) 16 registers.

35. _____ memory is used by the PLC for the operating system.
a) ROM
b) RAM
c) PROM
d) EEPROM

36. The most commonly used programming device is a:
a) personal computer.
b) dedicated industrial CRT terminal.
c) hand-held programmer.
d) miniprogrammer.

TEST 2•2

Place the answers to the following questions in the answer column at the right.

Answer

1. The I/O section of a PLC system usually consists of an I/O rack and individual I/O _____.

    1. _____

2. All I/O sections are always located next to the central processing unit. (True or False)

    2. _____

3. The location of a module within a rack and the terminal number of a module to which an input or output device is connected will determine the device's _____.

    3. _____

4. Most input modules have blown fuse indicators. (True or False)

    4. _____

5. The I/O address is used by the processor to identify where the device is _____.

    5. _____

6. A standard I/O module consists of a(n) (a) _____ board and a(n) (b) _____ assembly.

    6a. _____

    6b. _____

7. I/O modules are designed to plug into a slot or connector. (True or False)

    7. _____

8. Discrete I/O interfaces allow only _____-type devices to be connected.

    8. _____

9. I/O modules' circuity can be divided into two basic sections: the (a) _____ section and the (b) _____ section.

    9a. _____

    9b. _____

10. Optical isolation used in I/O modules helps reduce the effects of electrical noise. (True or False)

    10. _____

11. AC output modules often use a solid-state device such as a(n) _____ to switch the output ON and OFF.

    11. _____

12. The maximum current rating for the individual outputs of an ac output module is usually in the 20 to 30 ampere range. (True or False)

    12. _____

13. A(n) _____ relay is used for controlling larger load currents.

13. _____

14. Analog input interface modules contain a(n) _____ converter circuit.

14. _____

15. A thermocouple would be classified as an analog input sensing device. (True or False)

15. _____

16. _____ cable is often used in low voltage input signal circuits to reduce unwanted electrical noise.

16. _____

17. Electrical noise usually causes permanent operating errors. (True or False)

17. _____

18. Match each of the following specifications with the appropriate description. Place the number from the specifications list in the answer column.

SPECIFICATIONS
1) nominal current per input
2) ON-state input voltage range
3) OFF-state leakage current
4) electrical isolation
5) input delay
6) nominal input voltage
7) surge current
8) output voltage range
9) maximum output current rating
10) nominal output voltage

DESCRIPTIONS
a) Maximum voltage isolation between the I/O circuits and the controller logic circuitry.

18a. _____

b) Maximum value of current that flows through the output in its OFF state.

18b. _____

c) Maximum inrush current and duration an output module can withstand.

18c. _____

d) Maximum current that a single output and the module as a whole can safely carry.

18d. _____

e) Minimum and maximum output operating voltages.

18e. _____

f) Magnitude and type of voltage source that can be controlled by the output.

18f. _____

g) Duration for which the input must be ON before being recognized as a valid input.

18g. _____

h) Minimum input current that the input device must be capable of driving to operate the input circuit.

18h. _____

i) Voltage level at which the input signal is recognized as being ON.

18i. _____

j) Magnitude and type of voltage signal that will be accepted by the input.

18j. _____

19. The processor continually interacts with the _____ to interpret and execute the user program.

19. _____

20. The processor may perform functions such as timing, counting, and comparing in addition to logic processing. (True or False)

20. _____

21. Memory is where the control plan is held or stored in the controller. (True or False)

21. _____

22. Memory elements store individual pieces of information called _____.

22. _____

23. One K of memory capacity is the equivalent of 1024 words of memory storage. (True or False)

23. _____

24. A volatile memory will lose its programmed contents if operating power is lost. (True or False)

24. _____

25. A nonvolatile memory will retain its programmed contents if operating power is lost. (True or False)

25. _____

26. Match each of the following memory types with the operating characteristic.

MEMORY TYPES
1) PROM
2) EPROM
3) RAM
4) ROM
5) EEPROM

CHARACTERISTICS
a) Designed to permanently store a program that cannot be altered under any circumstances.

26a. _____

b) Is often referred to as read/write memory and is designed so that information can be written into or read from memory.

26b. _____

c) A special type of read-only memory that allows initial and/or additional information to be written into the chip only once after being received from the manufacturer.

26c. _____

d) Designed to be reprogrammed after being erased with ultraviolet light.

26d. _____

e) A special type of programmable read-only memory that can be reprogrammed after being electrically erased.

26e. _____

27. Today's controllers, for the most part, use magnetic-core memory. (True or False)

27. _____

28. Personal computers are the most commonly used PLC programming devices. (True or False)

28. _____

29. Analog signals can have an infinite number of states. (True or False)

29. _____

30. Electronic memory modules used for program loaders usually contain _____ memory.

30. _____

31. Identify each of the indicated parts of the I/O module shown in Fig. 2-8.

31a. _____

31b. _____

31c. _____

31d. _____

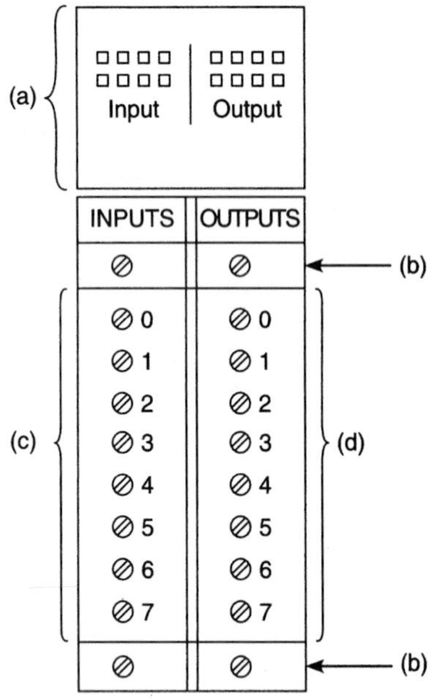

Fig. 2-8

32. A typical PLC has room for several I/O modules, which allows it to be customized for a particular application. (True or False)

32. _____

33. A logical rack is an I/O addressing unit that uses _____ words in the input and output image tables.

33. _____

34. There can be more than one rack in a chassis and more than one chassis in a rack. (True or False)

34. _____

35. Remote I/O racks are linked to the local rack through a(n) _____ module.

35. _____

36. In general, basic addressing elements include (a) _____, (b) _____, and (c) _____.

36a. _____

36b. _____

36c. _____

37. I/O modules are normally installed or removed while the PLC is powered. (True or False)

37. _____

38. A module inserted into the wrong slot could be damaged. (True or False)

38. _____

39. Modules receive voltage and current for proper operation from the _____ of the rack enclosure.

39. _____

40. Output modules can be purchased with (a) _____, (b) _____, or (c) _____ output.

40a. _____

40b. _____

40c. _____

41. The field device shown in Fig. 2-9 is a sinking device used with a(n) _____ output module.

41. _____

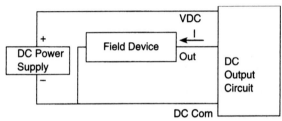

Fig. 2-9

42. The two basic types of analog input modules available are (a) _____ sensing and (b) _____ sensing.

42a. _____

42b. _____

43. Intelligent I/O modules have their own _____ on board.

43. _____

44. The CPU of a PLC system may contain more than one microprocessor. (True or False)

44. _____

45. Most PLC electronic components are not sensitive to electrostatic discharge. (True or False)

45. _____

46. Answer each of the following for the I/O module and status table shown in Fig. 2-10.

a) The type of module shown is a(n) _____ module.

b) The type of image table shown is a(n) _____ image table.

c) The status indicator associated with device *a* is _____. (ON or OFF)

d) The status indicator associated with device *b* is _____. (ON or OFF)

e) The value stored in memory for device *a* is _____.

f) The value stored in memory for device *b* is _____.

46a. _____

46b. _____

46c. _____

46d. _____

46e. _____

46f. _____

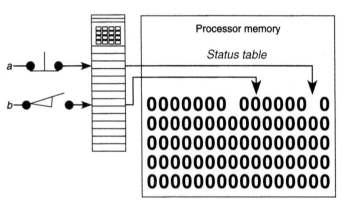

Fig. 2-10

47. Answer each of the following for the I/O module and status table shown in Fig. 2-11.

a) The type of module shown as a(n) _____ module.

b) The type of image table shown is a(n) _____ image table.

c) The status indicator associated with device *a* is _____. (ON or OFF)

d) The status indicator associated with device *b* is _____. (ON or OFF)

e) Device *a* is switched _____. (ON or OFF)

f) Device *b* is switched _____. (ON or OFF)

47a. _____

47b. _____

47c. _____

47d. _____

47e. _____

47f. _____

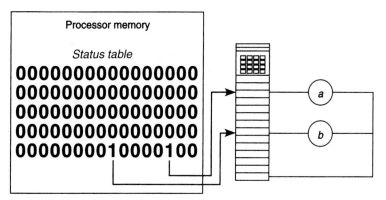

Fig. 2-11

48. Data are stored in memory locations by a process called _____.

48. _____

49. Data are retrieved from memory by what is referred to as _____.

49. _____

50. A UVPROM memory device requires battery backup. (True or False)

50. _____

51. Hand-held programmers allow you to document the PLC program. (True or False)

51. _____

52. Computer disk drives are most often used to record and store PLC programs. (True or False)

52. _____

53. The PLC can have only one program in memory at a time. (True or False)

53. _____

54. A PLC workstation screen is configured using personal computer-based setup software. (True or False)

54. _____

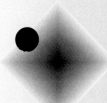

PROGRAMMING ASSIGNMENTS FOR CHAPTER 2

1. For the PLC you will be working with, record each of the following specifications for the input module(s):
   a) Inputs per module
   b) Nominal input voltage
   c) ON-state voltage range
   d) Nominal current per input
   e) Input delay
   f) Electrical isolation

2. For the PLC you will be working with, record each of the following specifications for the output module(s):
   a) Outputs per module
   b) Nominal output voltage
   c) Output voltage range
   d) Maximum output current rating per output
   e) Maximum output current rating per module
   f) Short circuit fuse protection
   g) Maximum surge current per output
   h) OFF-state leakage current per output
   i) Electrical isolation

3. For the PLC you will be working with, record each of the following specifications for the processor unit:
   a) Nominal input voltage
   b) Input voltage range
   c) Maximum power requirements
   d) Memory type
   e) User memory size
   f) I/O capacity
   g) Typical scan time
   h) Overload fuse protection

4. For the PLC you will be working with, identify all hardware components by means of a block diagram.

5. For the PLC you will be working with, document where the panel's input and output devices are wired.

6. Connect an ac input module with a single pushbutton input as shown in Fig. 2-7*b* of Chapter 2 of the text. Measure and record the amount of ac current that flows through the pushbutton when it is pressed.

7-1. Connect an ac output module with a single lamp load as shown in Fig. 2-9*b* of Chapter *2* of the text. Measure and record the amount of ac leakage current that flows through the lamp when this output is in the OFF state.

7-2. Program the controller to turn this output ON and record the amount of current that flows through the lamp when it is ON.

7-3. Measure and record the ac voltage across the lamp when the output is ON and OFF.

8. On a separate sheet of paper, draw a wiring layout for an interposing relay connection for the circuit of Question 6. Have this relay operate a secondary circuit of different voltage or current level using whatever relay and load are available.

9. Have your instructor provide you with detailed instructions for saving and restoring programs. Using the information provided, complete each of the following tasks:

a) Copy a program from a floppy disk to the programming terminal hard disk.

b) Restore a program from the hard disk to the processor memory.

c) Save a program from the processor memory onto the programming terminal hard disk.

d) Make a backup copy of a program from the programming terminal hard disk to a floppy disk.

10. Program the controller with two input switch devices and two output lamp loads as shown in Fig. 2-17 of Chapter 2 of the text. Enter the run mode and observe the input and output status table files.

3 NUMBER SYSTEMS AND CODES

TEST 3•1

Choose the letter that best completes the statement.   Answer

1. The decimal system has as its base: **1.** _____
   a) 2.   c) 8.
   b) 5.   d) 10.

2. Which of the following number systems has a base of 16?   **2.** _____
   a) Hexadecimal   c) Binary-coded decimal
   b) Octal   d) Gray code

3. In any number system, the position of a digit that represents part of   **3.** _____
   the number has a "weight" associated with its value. The place
   weights for binary:
   a) start with 1 and are successive powers of 2.
   b) increase by adding 2 for each place, starting with 0.
   c) increase by adding 2 for each place, starting with 2.
   d) start with 2 and double for each successive place.

4. The number 12 is: **4.** _____
   a) 12 in any number system.   c) 12 in binary.
   b) 12 in decimal.   d) all of the above.

5. The decimal number 15 would be written in binary as: **5.** _____
   a) 1111.   c) 4C.
   b) 1000.   d) 00011001.

6. The binary number 101 has the decimal equivalent of: **6.** _____
   a) 3.   c) 41.
   b) 101.   d) 5.

7. The number 127 could *not* be: **7.** _____
   a) decimal.   c) octal.
   b) hexadecimal.   d) binary.

8. The octal number 153 would be written in binary as: **8.** _____
   a) 011 101 001.   c) 011 111 101.
   b) 001 101 011.   d) 010 100 011.

9. The binary number 101101 would be written in decimal as: **9.** _____
   a) 21.   c) 45.
   b) 36.   d) 62.

10. The decimal number 28 would be written in binary as:
    a) 11100.   c) 10110.
    b) 00111.   d) 01011.

10. _____

11. The octal number 62 would be written in decimal as:
    a) A12.   c) 50.
    b) F35.   d) 98.

11. _____

12. The hexadecimal number C4 would be written in decimal as:
    a) 21.   c) 182.
    b) 48.   d) 196.

12. _____

13. The hexadecimal number 2D9 would be written in binary as:
    a) 0010 1101 1001.   c) 1100 1111 0010.
    b) 1001 1011 0010.   d) 0010 1011 1001.

13. _____

14. The decimal number 213 would be written in BCD as:
    a) 0010 0001 0011.   c) 0111 1001 0011.
    b) 1101 1000 1100.   d) 1011 1101 0101.

14. _____

15. A byte in Fig. 3-1 is represented by:
    a) No. 1.   c) No. 3.
    b) No. 2.   d) No. 4.

15. _____

No. 2 | 1 | 0 | 1 | 1 | 1 | 0 | 0 | 1 | 0 | 0 | 1 | 1 | 0 | 1 | 0 | 1 | No. 3

No. 1

No. 4

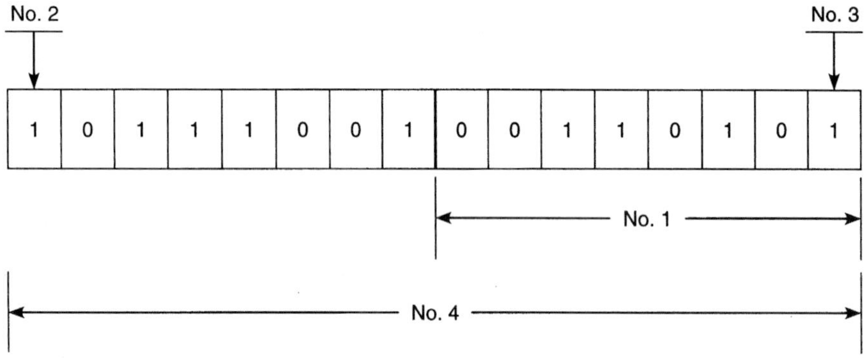

Fig. 3-1

16. The MSB in Fig. 3-1 is represented by:
    a) No. 1.   c) No. 3.
    b) No. 2.   d) No. 4.

16. _____

17. The main advantage of using the Gray code is:
    a) only one digit changes as the number increases.
    b) it can be easily converted to decimal numbers.
    c) large decimal numbers can be written using fewer digits.
    d) it uses the number 2 as its base.

17. _____

18. The acronym BCD stands for:
   a) binary-coded decimal.
   b) binary code decoder.
   c) base code decoder.
   d) base-coded decimal.

18. _____

19. For a base 8 number system, the *weight value* associated with the third digit would be:
   a) 16.
   c) 64.
   b) 32.
   d) 512.

19. _____

20. All digital computing devices operate using the binary number system because:
   a) most people are familiar with it.
   b) large decimal numbers can be represented in a shorter form.
   c) digital circuits can be easily distinguished between two voltage levels.
   d) all of the above.

20. _____

21. If a given memory unit consists of 1250 16-bit words, the memory capacity would be rated:
   a) 1250 bits.
   c) 3260 bits.
   b) 20,000 bits.
   d) 156 bits.

21. _____

22. In the sign bit position, a 1 indicates a(n):
   a) negative number.
   c) octal code.
   b) positive number.
   d) hexadecimal code.

22. _____

23. The 2's complement form of a binary number is the binary number that results when:
   a) all the 1's are changed to 0's.
   b) all the 0's are changed to 1's.
   c) 1 is added to the 1's complement.
   d) both *a* and *b*

23. _____

24. The ASCII code:
   a) is used with absolute encoders.
   b) is considered to be an error-minimizing code.
   c) includes letters as well as numbers.
   d) all of the above.

24. _____

25. A(n) _____ bit is used to detect errors that may occur while a word is moved.
   a) parity
   c) positive
   b) negative
   d) overflow

25. _____

TEST 3•2

Place the answers to the following questions in the answer column at the right.

Answer

1. PLCs work on _____ numbers in one form or another to represent various codes or quantities.

1. _____

2. The decimal system uses the number 9 as its base. (True or False)

2. _____

3. The only allowable digits in the binary system are (a) _____ and (b) _____.

3a. _____

3b. _____

4. Each digit of a binary number is known as a(n) _____.

4. _____

5. With reference to processor memory locations, the term *register* is often used interchangeably with _____.

5. _____

6. All digital computing devices perform operations in binary. (True and False)

6. _____

7. The base of a number system determines the total number of unique symbols used by that system. (True or False)

7. _____

8. Match the following bases with the appropriate number system.

8a. _____

Bases	Number Systems
1) Base 2	a) Binary
2) Base 16	b) Decimal
3) Base 10	c) Octal
4) Base 8	d) Hexadecimal

8b. _____

8c. _____

8d. _____

9. In any number system, the position of a digit that represents part of the number has a weighted value associated with it. (True or False)

9. _____

10. Match the following decimal numbers with their binary equivalent.

Decimal Numbers	Binary Equivalent
1) 9	a) 110011
2) 37	b) 1001
3) 51	c) 100101
4) 42	d) 101010

10a. _____

10b. _____

10c. _____

10d. _____

11. Usually a group of 8 bits is a byte, and a group of one or more bytes is a word. (True or False)

11. _____

12. The _____ _____ bit of a word is the digit that represents the smallest value.

12. _____

13. A memory that has a capacity of 700 sixteen-bit words can actually store _____ bits of information.

13. _____

14. To express a number in binary requires fewer digits than in the decimal system. (True or False)

14. _____

15. The octal number system consists of digits 0, 1, 2, 3, 4, 5, 6, and 7. There are no 8's or 9's. (True or False)

15. _____

16. The octal number 46 expressed as a decimal number would be _____.

16. _____

17. The octal number 153 expressed as a binary number would be _____.

17. _____

18. The hexadecimal number system consists of 16 digits including the numbers 0 through 9 and letters A through F. (True or False)

18. _____

19. Hexadecimal 2F equals _____ in decimal.

19. _____

20. Hexadecimal A6 equals _____ in binary.

20. _____

21. The decimal number 29 equals (a) _____ in binary and (b) _____ in BCD.

21a. _____

21b. _____

22. The BCD number 1000 0101 0110 0111 equals _____ in decimal.

22. _____

23. In the Gray code there is a maximum of one bit change between two consecutive numbers. (True or False)

23. _____

24. The radix of a number system is the same as the base. (True or False)

24. _____

25. Binary number systems use positive and negative symbols to represent the polarity of a number. (True or False)

25. _____

26. Two systems of parity are normally used: _____ and _____.

26a. _____

26b. _____

27. Add binary 11101 and 1100.

27. _____

28. Subtract binary 11101 from 111010.

28. _____

29. Multiply binary 110 and 111.

29. _____

30. Divide binary 11010 by 10.

30. _____

31. The three outputs of a four-bit A-B comparator would be (a) _____, (b) _____, and (c) _____.

31a. _____

31b. _____

31c. _____

PROGRAMMING ASSIGNMENTS FOR CHAPTER 3

2-1. Complete the following chart using the change radix function of the PLC or a scientific calculator.

Binary	Octal	Decimal	Hexadecimal
101			5
	11		
		15	
			D
	16		
1001011			
	47		
		73	

4 FUNDAMENTALS OF LOGIC

TEST 4•1

Choose the letter that best completes the statement.   Answer

1. The binary concept makes use of the fact that certain information: **1.** _____
   a) can exist in one of two possible states.
   b) can be broken down into smaller units for easier analysis.
   c) can be divided into two or more categories.
   d) can be divided into two, or multiples of two, categories.

2. A gate is a device that: **2.** _____
   a) allows current flow in one direction only.
   b) changes alternating current to direct current.
   c) performs a logical decision based on its inputs.
   d) performs a logical decision based on its outputs.

3. In conventional logic circuits, binary 1 represents: **3.** _____
   a) the presence of a signal.   c) a high voltage level.
   b) the occurrence of some event.   d) all of the above.

4. The logic function(s) used by PLCs is (are):   **4.** _____
   a) AND   c) NOT
   b) OR   d) all of the above.

5. The basic rule for an AND gate is: **5.** _____
   a) if all inputs are 1, the output will be 1.
   b) if all inputs are 1, the output will be 0.
   c) if all inputs are 0, the output will be 1.
   d) both *a* and *b.*

6. The basic rule for an OR gate is: **6.** _____
   a) if one or more inputs are 1, the output is 1.
   b) if one or more inputs are 1, output is 0.
   c) if one or more inputs are 0, the output is 1.
   d) both *b* and *c.*

7. The NOT function can be thought of as: **7.** _____
   a) a FALSE-to-TRUE converter.   c) an inverter.
   b) a changer of states.   d) all of the above.

8. A NOT function is used when a logic 1 must _____ some device.   **8.** _____
a) activate   c) switch
b) deactivate   d) light

9. The OR function, implemented using contacts, requires contacts   **9.** _____
connected in:
a) series   c) series/parallel
b) parallel   d) parallel/series

10-1. The logic symbol drawn in Fig. 4-1 is that of a(n):   **10-1.** _____
a) AND function.   c) NOT function.
b) OR function.   d) NAND function.

10-2. The Boolean equation for the logic symbol drawn in Fig. 4-1 is: **10-2.** _____
a) $Y = A + B$ b) $Y = AB$
c) $Y = A \cdot B$ d) either *b* or *c*.

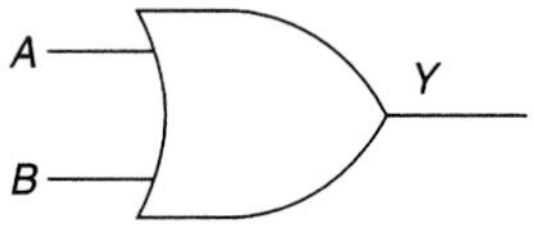

```
A ───┐
     ) ── Y
B ───┘
```

Fig. 4-1

11-1. The logic symbol drawn in Fig. 4-2 is that of a(n):   **11-1.** _____
a) AND function.   c) NOT function.
b) OR function.   d) NOR function.

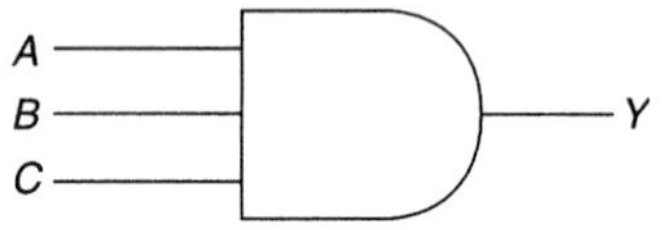

```
A ───┐
B ───┤── Y
C ───┘
```

Fig. 4-2

11-2. The Boolean equation for the logic symbol drawn in Fig. 4-2 is: **11-2.** _____
a) $Y = A + B + C$ c) $Y = (AB) + C$
b) $Y = ABC$ d) $Y = (A - B)C$

12. With reference to the logic circuit of Fig. 4-3, the output *Y* will be at   **12.** _____
a logic 1 when:
a) inputs *A* and *B* are logic 1.
b) input *A* or *B* is logic 1.
c) input *A* is at logic 1 and input *B* is at logic 0.
d) input *A* is at logic 0 and input *B* is at logic 1.

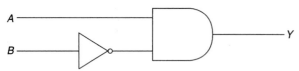

Fig. 4-3

13-1. The logic symbol drawn in Fig. 4-4 is that of a(n):
  a) AND function.   c) NOR function.
  b) OR function.   d) NAND function.

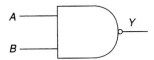

Fig. 4-4

13-2. The truth table for the logic symbol drawn in Fig. 4-4 is:

Inputs	Output
A B	Y
0 0	1
0 1	0
1 0	0
1 1	0

(a)

Inputs	Output
A B	Y
0 0	0
0 1	1
1 0	1
1 1	1

(c)

Inputs	Output
A B	Y
0 0	1
0 1	1
1 0	1
1 1	0

(b)

Inputs	Output
A B	Y
0 0	0
0 1	0
1 0	0
1 1	1

(d)

14-1. The logic symbol drawn in Fig. 4-5 is that of a(n):
  a) NOT function.   c) NAND function.
  b) NOR function.   d) OR function.

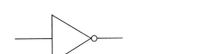

Fig. 4-5

14-2. The Boolean equation for the logic symbol drawn in Fig. 4-5 is:
  a) $A = Y$ c) $\overline{A} = \overline{Y}$
  b) $A = B$ d) $A = \overline{Y}$

15. The Boolean expression for the logic circuit drawn in Fig. 4-6 is:
a) $Y = ABC$ c) $Y = (A + B)C$
b) $Y = A + B + C$ d) $Y = AB + C$

15. _____

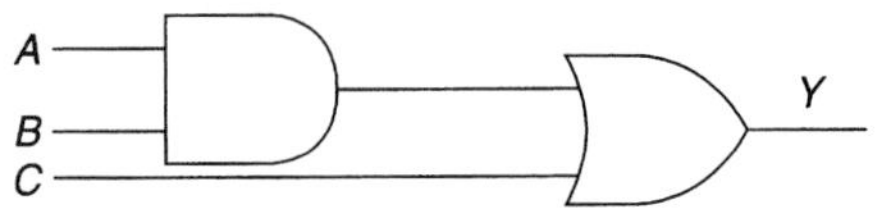

Fig. 4-6

16. The Boolean expression for the logic circuit drawn in Fig. 4-7 is:
a) $Y = (A B)(CD)$
b) $Y = (\overline{A + B + C})D$
c) $Y = \overline{ABC} + D$
d) $Y = ABC + D$

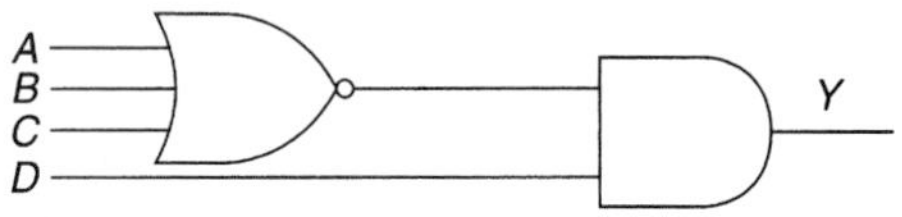

16. _____

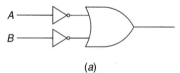
Fig. 4-7

17. Which of the following algebraic laws holds true for Boolean algebra only?
a) $A + B = B + A$
b) $A \cdot B = B \cdot A$
c) $A \cdot (B + C) = (A \cdot B) + (A \cdot C)$
d) $A + (B \cdot C) = (A + B) \cdot (A + C)$

17. _____

18. According to De Morgan's law, the logic circuit of Fig. 4-8 is equivalent to:

18. _____

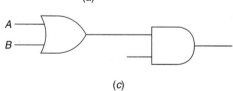

(a)

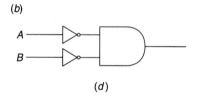

(b)

(c)

(d)

Fig. 4-8

19. Which of the following is the circuit for the Boolean expression
$Y = A(BC + D)$:

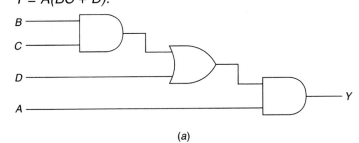

(a)

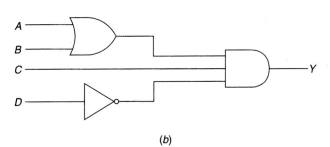

(b)

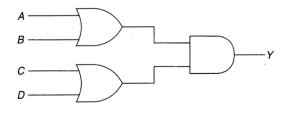

(c)

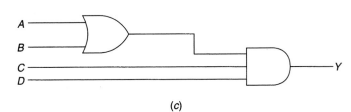

(d)

20. Which of the following is the circuit for the Boolean expression
$Y = (A + B)(\bar{C} + D)$?

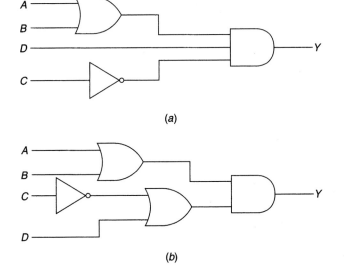

(a)

(b)

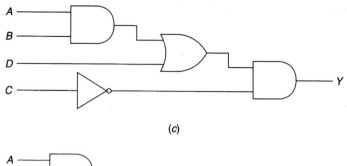

(c)

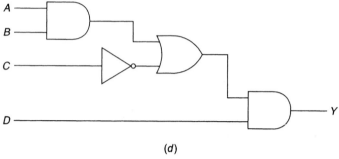

(d)

(Continued from page 45, Question 20)

21. The Boolean expression for the logic circuit drawn in Fig. 4-9 is:

    a) $Y = \bar{A}\bar{B} + AB$ c) $Y = \bar{A}B + A\bar{B}$

    b) $Y = (\bar{A}B)(A\bar{B})$ d) $Y = (\bar{A} + B)(A + \bar{B})$

21. _____

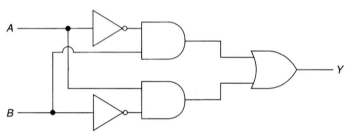

Fig. 4-9

22. The Boolean expression for the logic circuit drawn in Fig. 4-10 is:

    a) $(A + E)\,DC\bar{B}$ c) $(AE + D)\,C\bar{B}$

    b) $AE(\bar{D} + C + \bar{B})$ d) $(A + E)\,DC\bar{B}$

22. _____

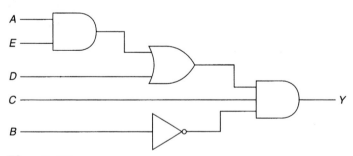

Fig. 4-10

23-1. With reference to the ladder logic program of Fig. 4-11, part No. 1 represents the:

    a) indicating light.   c) normally open contact.

    b) normally closed contact.   d) relay coil equivalent.

23-1. _____ *d*

23-2. With reference to the ladder logic program of Fig. 4-11, part No. 2 represents the:

  a) rail.   c) + side of voltage source.
  b) rung.   d) − side of voltage source.

23-2. _a_

23-3. With reference to the ladder logic program of Fig. 4-11, the number of outputs per rung is:

  a) 1   c) 3
  b) 2   d) unlimited.

23-3. _A_

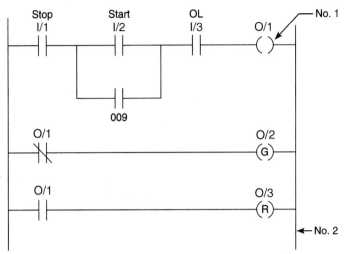

Fig. 4-11

24-1. Which of the following is the relay schematic for the ladder log program drawn in Fig. 4-12?

24-1. _b_

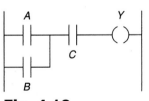

Fig. 4-12

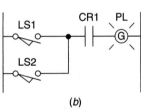

(a)

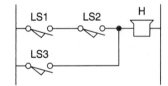

(c)

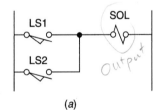

(b)

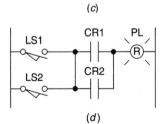

(d)

24-2. Which of the following is the equivalent logic circuit symbol for the ladder logic program drawn in Fig. 4-12?

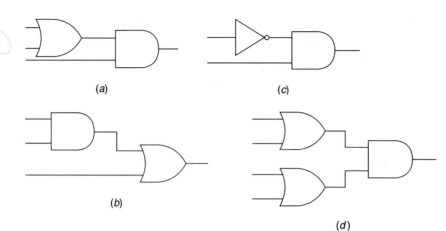

(a)

(c)

(b)

(d)

24-3. Which of the following is the Boolean equation for the ladder logic program drawn in Fig. 4-12?

a) $(AB) + C = Y$ c) $ABC = Y$
b) $A + B + C = Y$ d) $(A + B)C = Y$

25. A single rung of a ladder logic program is arranged with:

   a) input conditions connected from left to right, with the output at the far right.
   b) input conditions connected from right to left, with the output at the far left.
   c) the output in the center and the input conditions to the left and right of it.
   d) all input conditions in parallel and all output conditions in series.

26. An AND gate operates on the same principle as:

   a) a series circuit.   c) a series-parallel circuit.
   b) a parallel circuit.   d) none of the above.

27. A NOR gate is:

   a) an AND gate with an inverter connected to the output.
   b) an OR gate with an inverter connected to the output.
   c) equivalent to a series circuit.
   d) equivalent to a parallel circuit.

28. The basic rule for an XOR function is:

   a) if one or the other, but not both, inputs are 1, the output is 1.
   b) if one or more inputs are 1, the output is 1.
   c) if one or more inputs are 1, the output is 0.
   d) if one or more inputs are 0, the output is 1.

29. If you want to know when one or both matching bits in two different words are ON, you would use the _____ logic instruction.

   a) AND   c) OR
   b) XOR   d) NOT

TEST 4•2

Place the answers to the following questions in the answer column at the right.

Answer

1. The binary concept used in logic refers to the fact that many things can be thought of as existing in one of _____ states.

1. _____

2. Normally, a binary 1 represents the presence of a signal, while a binary 0 represents the absence of a signal. (True or False)

2. _____

3. A light that is ON or a switch that is closed would normally be represented by a binary _____.

3. _____

4. All gates are devices that have one input with which they perform logic decisions and produce a result at one or more of their outputs. (True or False)

4. _____

5. The _____ gate output is 1 only if all inputs are 1.

5. _____

6. The _____ gate output is 1 if one or more of its inputs are 1.

6. _____

7. The NOT output is 1 if the input is _____.

7. _____

8. The NOT function is also called a(n) _____.

8. _____

9. Match the symbols in Fig. 4-13 with their correct logic function. Place the number from the symbols in Fig. 4-13 in the answer column.

SYMBOLS

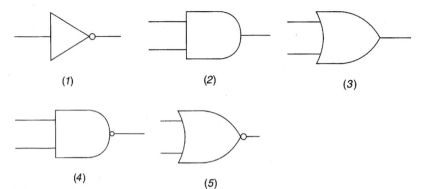

(1) (2) (3)

(4) (5)

Fig. 4-13

LOGIC FUNCTIONS
a) AND
b) OR
c) NOR
d) NOT
e) NAND

9a. _____

9b. _____

9c. _____

9d. _____

9e. _____

10. The mathematical study of the binary number system and logic is called _____ algebra.

10. _____

11. The AND function, implemented using switches, will mean switches connected in parallel. (True or False)

11. _____

12. The OR function, expressed as a Boolean equation, would be $Y = AB$. (True or False)

12. _____

13. The correct Boolean expression for the logic circuit in Fig. 4-14 is _____.

13. _____

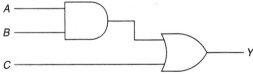

Fig. 4-14

14. The correct Boolean expression for the logic circuit in Fig. 4-15 is _____.

14. _____

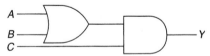

Fig. 4-15

15. The correct Boolean expression for the logic circuit in Fig. 4-16 is _____.

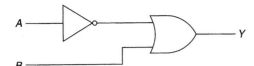

Fig. 4-16

16. The correct Boolean expression for the logic circuit in Fig. 4-17 is _____.

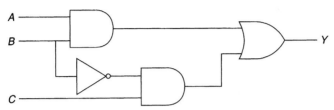

Fig. 4-17

17. *Hardwired logic* refers to logic control functions determined by the way devices are interconnected. (True or False)

18. Hardwired logic can be implemented using relays and relay schematics. (True or False)

19. Hardwired logic is fixed and is changeable only by altering the way devices are connected. (True or False)

20. Programmable control is based on logic functions that are programmable and easily changed. (True or False)

21. _____ symbology is a way of expressing the PLC control logic in terms of relay-equivalent contacts and coils.

22. There is no difference between a relay schematic and a ladder logic program. (True or False)

23. On some PLCs, only one output is allowed per ladder logic rung. (True or False)

24. A rung is the contact symbology required to control an output. (True or False)

25. The most common PLC _____ is ladder logic programs.

15. _____

16. _____

17. _____

18. _____

19. _____

20. _____

21. _____

22. _____

23. _____

24. _____

25. _____

26. Complete the truth table of basic Boolean operations by signifying the correct true or false condition for each blank space.

26. _____

A	B	A and B	A or B	not A	A xor B
False	False				
False	True				
True	False				
True	True				

27. What will be the data stored in the destination address B3:4 of Fig. 4-18 when the input is true?

27. _____

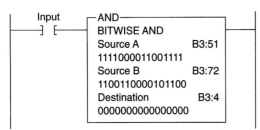

Fig. 4-18

28. What will be the data stored in the destination address B3:27 of Fig. 4-19 when the input is true?

28. _____

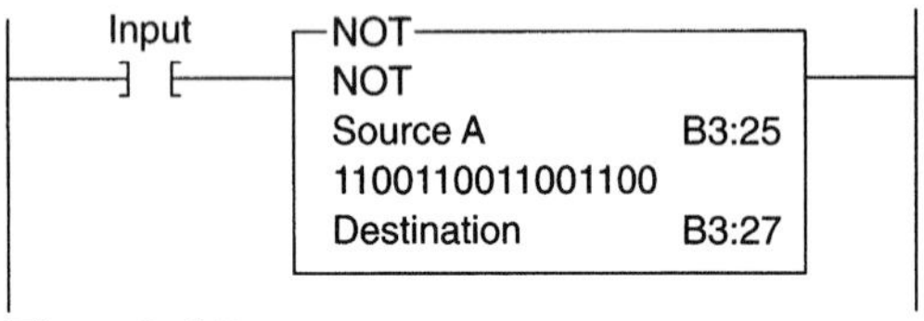

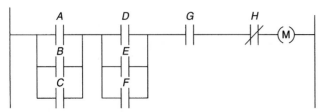

Fig. 4-19

29. Draw a PLC ladder diagram program for the gate logic array shown in Fig. 4-20.

29. _____

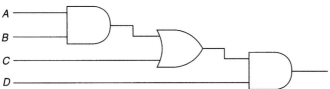

Fig. 4-20

30. Draw the equivalent gate logic array for the PLC ladder diagram shown in Fig. 4-21.

30. _____

Fig. 4-21

PROGRAMMING ASSIGNMENTS FOR CHAPTER 4

1. Program the relay schematic of Fig. 4-22 using a PLC and check for operation.

Relay schematic   Ladder logic program

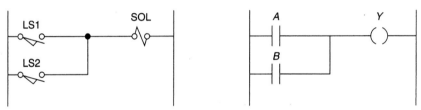

Boolean equation: $AB = Y$

Fig. 4-22

2. Program the relay schematic of Fig. 4-23 using a PLC and check for operation.

Relay schematic   Ladder logic program

Boolean equation: $A + B = Y$

Fig. 4-23

3. Program the relay schematic of Fig. 4-24 using a PLC and check for operation.

Relay schematic   Ladder logic program

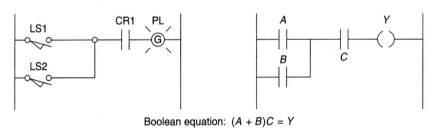

Boolean equation: $(A + B)C = Y$

Fig. 4-24

4. Program the relay schematic of Fig. 4-25 using a PLC and check for operation.

5. Program the relay schematic of Fig. 4-26 using a PLC and check for operation.

6. Program the relay schematic of Fig. 4-27 using a PLC and check for operation.

7. Program the relay schematic of Fig. 4-28 using a PLC and check for operation.

8. Express each of the following equations as a ladder logic rung. Program each rung into the PLC and prove its operation.
a) $Y = (A + B)CD$
b) $Y = (A\bar{B}C) + \bar{D} + E$
c) $Y = [(\bar{A} + \bar{B})C] + D\bar{E}$
d) $Y = (AB\bar{C}) + (D\bar{E}F)$

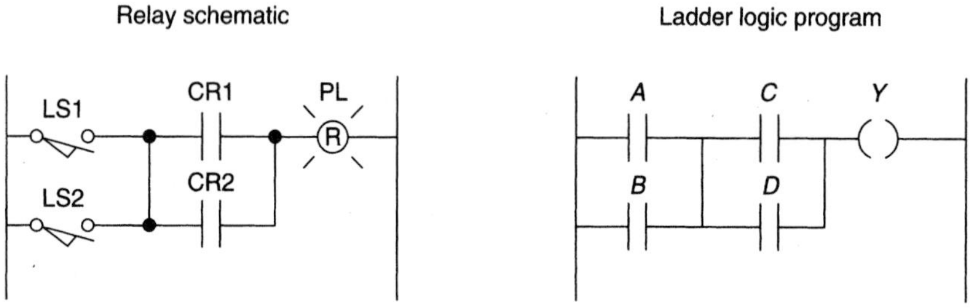

Relay schematic   Ladder logic program

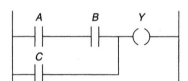

Boolean equation: $(A + B)(C + D) = Y$

Fig. 4-25

Relay schematic   Ladder logic program

Boolean equation: $(AB) + C = Y$

Fig. 4-26

Relay schematic   Ladder logic program

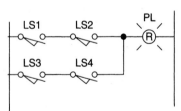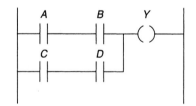

Boolean equation: $(AB) + (CD) = Y$

Fig. 4-27

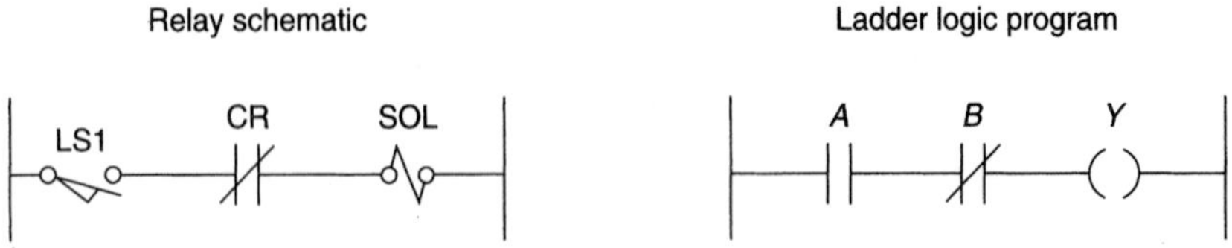

Relay schematic   Ladder logic program

Boolean equation: $A\bar{B} = Y$

Fig. 4-28

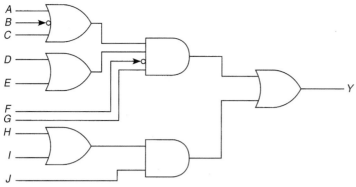

Fig. 4-29

9. Develop a PLC program that will simulate the operation of the XOR function. Enter the program into the PLC and prove its operation.

10. A conveyor will run when any one of four inputs are on. It will stop when any one of four other inputs are on. Develop a PLC program that will simulate this operation. Enter the simulated program into the PLC and prove its operation.

11. Develop a PLC program that will simulate the gate array logic shown in Fig. 4-29. Enter the simulated program into the PLC and prove its operation.

12. Enter the logical AND program shown in Fig. 4-30 into the PLC. Use the data monitor function to store the following data:

    B3:1 = 1111 0000 1111 0000

    B3:2 = 0000 0000 1111 0000

    Run the program and verify that B3:3 contains the following bit pattern:

    B3:3 = 0000 0000 1111 0000

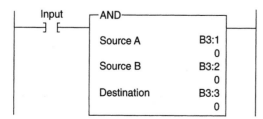

Fig. 4-30

13. Enter the logical XOR program shown in Fig. 4-31 into the PLC. Use the data monitor function to store the following data:

    B3:5 = 1010 1100 0111 1111

    B3:6 = 1010 0111 0111 0111

    Run the program and verify that B3:7 contains the following bit pattern:

    B3:7 = 0000 1011 0000 1000

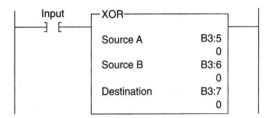

Fig. 4-31

14. Assume you have an alarm condition as part of a machine operation that tells you when one (or more) of eight limit switches wired into one input module is not in the correct position to allow the process to continue. Develop a program that uses the XOR logical instruction and that provides a simple means for the troubleshooter to isolate which switch (or switches) is in the wrong position by operating a pushbutton. Enter the program into the PLC and prove its operation.

5 BASICS OF PLC PROGRAMMING

TEST 5•1

Choose the letter that best completes the statement.

Answer

1. The _____ will account for most of the total memory of a given PLC system.
   a) input image table file   c) user program
   b) output image table file   d) internal operating instructions

   1. C

2. The status bit of switches and pushbuttons connected to a PLC is stored in the:
   a) input image table file.   c) user program.
   b) output image table file.   d) all of the above.

   2. A

3-1. The module shown in Fig. 5-1 is:
   a) an input module.   c) an output module.
   b) a programming module.   d) a memory module.

   3-1. C

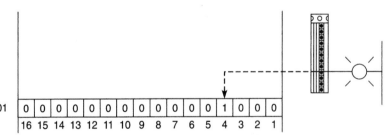

001	0	0	0	0	0	0	0	0	0	0	0	0	1	0	0	0
	16	15	14	13	12	11	10	9	8	7	6	5	4	3	2	1

Fig. 5-1

3-2. The status bit of the light shown in Fig. 5-1 is stored in:
   a) word 001 and bit 04.   c) word 001 1000.
   b) word 04 and bit 001.   d) word 0001 001.

   3-2. A

3-3. The status bit of the light shown in Fig. 5-1 is controlled by the:
   a) connection of outputs.
   b) connection of inputs.
   c) processor as it interprets the user program.
   d) EXAMINE IF CLOSED instruction.

   3-3. B

4. With reference to the typical program scan illustrated in Fig. 5-2, part A represents the:
  a) program scan.
  b) period during which the instructions are executed.
  c) I/O scan.
  d) both *a* and *b.*

4. _____C_____

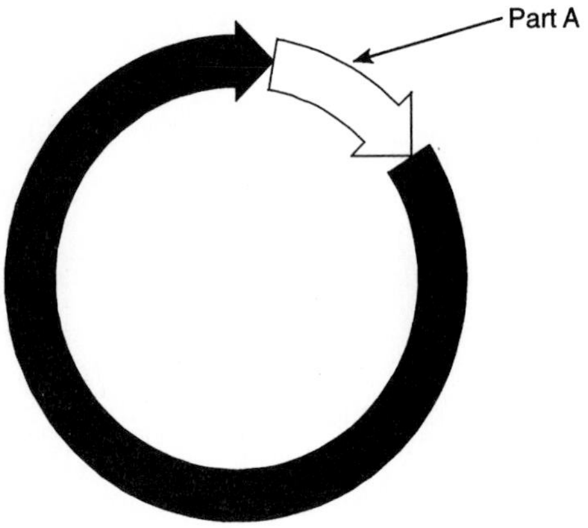

— Part A

Fig. 5-2

5. *Memory organization* defines how certain areas of memory are used and is commonly referred to as a:
  a) program.
  b) data sheet.
  c) processor.
  d) memory map.

5. _____d_____

6-1. If a PLC has a total scan time of 10 ms and has to monitor a signal that _____, then the controller may not detect this change.
  a) changes state once in 20 ms   c) is constantly changing
  b) is fast   d) changes state twice in 5 ms

6-1. _____d_____

6-2. Which of the following is *not* a factor in determining the total scan time of a PLC?
  a) Memory protection
  b) Program content
  c) Program length
  d) Total number of inputs and outputs used

6-2. _____A_____

6-3. The scan is normally a sequential process of:
  a) reading the control logic, evaluating the outputs, and updating the inputs.
  b) writing the control logic, evaluating the outputs, and updating the inputs.
  c) reading/writing the status of inputs and updating the outputs.
  d) reading the status of inputs, evaluating the control logic, and updating the outputs.

6.3. _____D_____

7-1. The PLC programming language refers to the method by which:
   a) the user program is executed by the processor.
   b) outputs are updated by the PLC.
   c) the user programs the PLC.
   d) the memory of the PLC is organized by the manufacturer.

7-1. ~~A~~ C

7-2. The most common PLC language is:
   a) ladder diagram.   c) English statements.
   b) Boolean statements.   d) binary-coded decimal.

7-2. A

8-1. In Fig. 5-3, *A* represents an:
   a) EXAMINE IF CLOSED instruction.
   b) EXAMINE IF OPEN instruction.
   c) OUTPUT ENERGIZE instruction.
   d) INPUT ENERGIZE instruction.

8-1. A

8-2. In Fig. 5-3, *B* represents an:
   a) EXAMINE IF CLOSED instruction.
   b) EXAMINE IF OPEN instruction.
   c) OUTPUT ENERGIZE instruction.
   d) INPUT ENERGIZE instruction.

8-2. B

8-3. In Fig. 5-3, *Y* represents an:
   a) EXAMINE IF CLOSED instruction.
   b) EXAMINE IF OPEN instruction.
   c) OUTPUT ENERGIZE instruction.
   d) INPUT ENERGIZE instruction.

8-3. C

8-4. In Fig. 5-3, which of the following input combinations will result in an output?
   a) *A* and *B* and *C* and *D*   c) *E* and *D*
   b) *A* and not *B* and *C* and *D*   d) *E* and not *D*

8-4. C

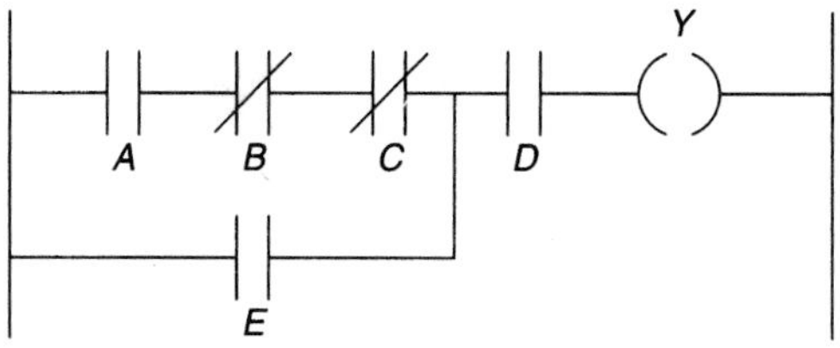

Fig. 5-3

9. A programmed normally closed PLC contact:
   a) with a bit status of 1 will not have logic continuity.
   b) is examined for an OFF condition.
   c) is examined for an ON condition.
   d) both *a* and *b*.

9. D

10. The programming of a normally open PLC contact is called for when the:
   a) normal state of the input device is ON.
   b) normal state of the input device is OFF.
   c) contacts must close to energize the output.
   d) contacts must open to energize the output.

10. ~~C~~ B

11. A ladder rung is said to be TRUE, or have logic continuity, when:
    a) its output is energized.
    b) at least one left-to-right path of contacts is closed.
    c) all contacts are at a logic 1.
    d) both *a* and *b.*

11. _____

12. The most likely five-digit address for LS1 of Fig. 5-4 would be:
    a) O:01/02.   c) I:02/01.
    b) LS1 01.   d) LS1 02.

12. _____

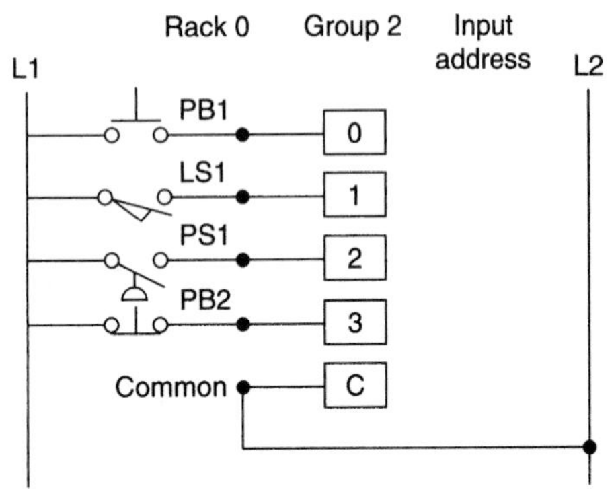

Fig. 5-4

13. The addressing format used with PLCs:
    a) is always a five-digit code.
    b) is always represented using the octal number system.
    c) is standard for all manufacturers.
    d) none of the above.

13. _____

14. Branch instructions are used to create:
    a) series paths.
    b) parallel paths.
    c) a series of energized output conditions.
    d) all of the above.

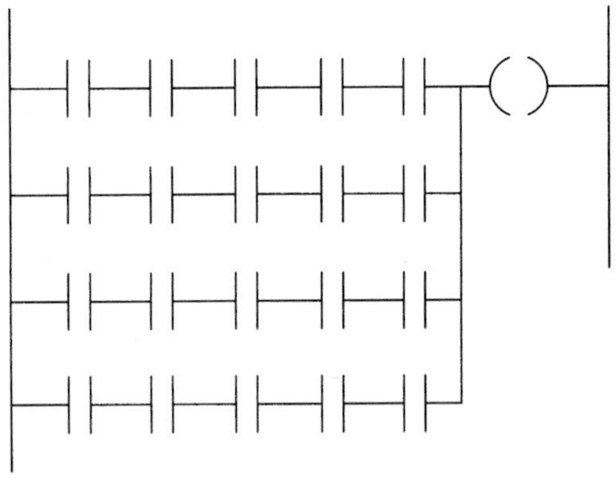

14. _____

Fig. 5-5

15-1. In the matrix limitation diagram of Fig. 5-5, the maximum number of parallel rows allowed is:
    a) one.   c) five.
    b) four.   d) limited only by the memory size.

15-1. _____

15-2. The maximum number of rungs allowed is:
    a) one.
    b) four.
    c) five.
    d) limited only by the memory size.

15-2. _____

15-3. The maximum number of series contacts allowed per rung is:
    a) one.
    b) four.
    c) five.
    d) limited only by the memory size.

15-3. _____

15-4. The maximum number of outputs allowed per rung is:
    a) one.
    b) four.
    c) five.
    d) limited only by the memory size.

15-4. _____

16. In Fig. 5-6, the *nested* contact is contact:
    a) *B.*
    b) *C.*
    c) *D.*
    d) *E.*

16. _____

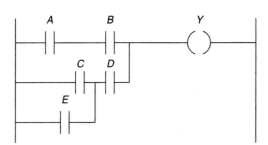

Fig. 5-6

17. The Boolean equation for the logic represented in the ladder logic program in Fig. 5-7 can be expressed as:
    a) $Y = (A) + (CD) + (BC)$
    b) $Y = A + B + (CD)$
    c) $Y = (AB) + (CD)$
    d) $Y = A(BCD)$

17. _____

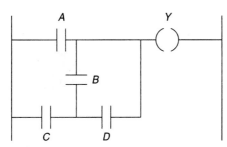

Fig. 5-7

18. The ladder logic program in Fig. 5-7 can be reprogrammed as shown in Fig. 5-8, in _____, to eliminate the vertical programmed contact and maintain the same input logic conditions.

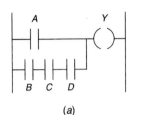

(a)

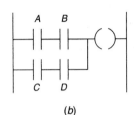

(b)

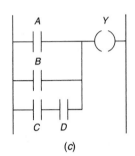

(c)

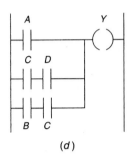

(d)

Fig. 5-8

19. With reference to the ladder logic program of Fig. 5-9, if it could be programmed as shown, part of the logic would be ignored due to the fact that the processor allows for a flow:
a) from right to left only.
b) from left to right only.
c) in the upward direction only.
d) both *a* and *c*.

20. The Boolean equation for the logic represented in the ladder diagram of Fig. 5-9 can be expressed as:
a) $Y = (AB) + (ACD) + (DE)$
b) $Y = (AB) + (ACD) + (DE) + (BCE)$
c) $Y = (AB) + (AC) + (AD) + (ED)$
d) $Y = (AB) + (CD) + E$

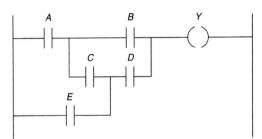

Fig. 5-9

21. An internal control relay:
a) does not directly control an output device.
b) is not controlled by the programmed logic.
c) is used primarily to control the internal power to the processor module.
d) is used to control multiple output circuits.

22. In Fig. 5-10, the bit status condition of the input device connected to address 101/14 must be _____ to turn on output address 001/04.
    a) 0
    b) 1
    c) normally open
    d) normally closed

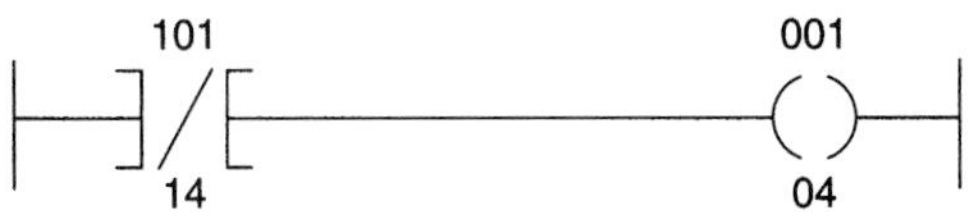

22. _____

```
     101                        001
 ├──┤/├──────────────────────( )──┤
     14                         04
```

Fig. 5-10

23. For most PLC systems, one word of memory is required for each:
    a) EXAMINE IF CLOSED and EXAMINE IF OPEN instruction.
    b) OUTPUT ENERGIZE instruction.
    c) branch START/END instruction.
    d) all of the above.

23. _____

24. When the PLC is required to operate the user program without energizing any outputs, it is placed in the _____ mode.
    a) RUN
    b) CLEAR MEMORY
    c) PROGRAM
    d) TEST

24. _____

25-1. In ladder logic programs, outputs are represented by:
    a) contact symbols.
    b) coil symbols.
    c) schematic load device symbols.
    d) either *a* or *b*.

25-1. _____

25-2. Assume that a program rung consists of an OUTPUT ENERGIZE instruction only. The output would then be:
    a) continuously ON.
    b) continuously OFF.
    c) shorted.
    d) both *a* and *c*.

25-2. _____

26. Parallel connections of a ladder logic program are typically called:
    a) rungs.
    b) networks.
    c) coils.
    d) branches.

26. _____

27. Each complete horizontal line of a ladder diagram is generally referred to as a(n):
    a) rung.
    b) branch.
    c) input.
    d) output.

27. _____

28. The scan time of a PLC is equal to:
    a) the time taken to scan the user program.
    b) the time taken to scan the inputs and outputs.
    c) the speed at which field contacts close and outputs can be operated.
    d) the time taken to scan inputs and outputs and execute the user program.

28. _____

29. The last element to be entered on a ladder logic rung is a(n):
    a) coil.
    b) contact.
    c) EXAMINE IF CLOSED.
    d) EXAMINE IF OPEN.

29. _____

30. A(n) _____ scan pattern examines instructions rung by rung.
    a) horizontal
    b) immediate
    c) input
    d) output

30. _____

31. Function chart programming allows:
    a) faster scan time.
    b) simpler programming.
    c) simplified troubleshooting.
    d) all of the above.

31. _____

TEST 5•2

Place the answers to the following questions in the answer column at the right.

Answer

1. All PLC manufacturers organize their memories in the same way. (True or False)

1. _____

2. The memory organization of a PLC is often called a memory _____.

2. _____

3. The memory space can be divided into the two broad categories of (a) _____ and (b) _____.

3a. _____

3b. _____

4. The _____ portion of the memory is where the programmed logic ladder program is entered and stored.

4. _____

5. Most of the total PLC memory is used for the _____.

5. _____

6. The user program contains the logic that controls the machine operation. (True or False)

6. _____

7. Most instructions require one word of memory. (True or False)

7. _____

8. The status of input and output devices is stored in the data table. (True or False)

8. _____

9. If a switch connected to an input module is closed, a binary 1 is stored in the proper _____ location.

9. _____

10. The _____ image table file is updated during the I/O scan to reflect the current status of digital inputs.

10. _____

11. If the program calls for a specific output to be ON, the corresponding bit in the output image table file is set to _____.

11. _____

12. The _____ is normally a continuous and sequential process of reading the status of inputs, evaluating the control logic, and updating the outputs.

12. _____

13. The PLC programming _____ refers to the method by which the user communicates the information to the PLC.

13. _____

14. The greater the scan time, the faster the PLC can react to changes in inputs. (True or False)

14. _____

15. Scan time varies with program content and length. (True or False)

15. _____

16. If any input signal changes state very quickly, it is possible that the controller may never be able to detect the change. (True or False)

16. _____

17. The most common PLC programming language is BASIC. (True or False)

17. _____

18. The ladder logic program language is basically a(n) _____ set of instructions used to create the controller program.

18. _____

19. (a) _____ (b) _____ are the basic symbols of the ladder logic program instruction set.

19a. _____

19b. _____

20. Match the symbols in Fig. 5-11 with their correct instruction. Place the number from the symbols list in the answer column.

SYMBOLS

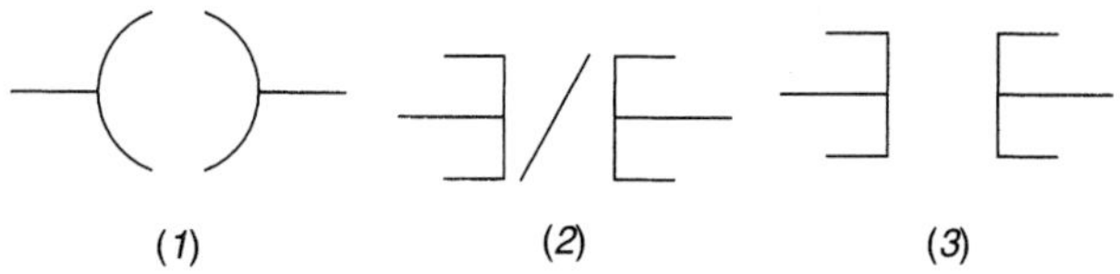

(1)   (2)   (3)

Fig. 5-11

INSTRUCTIONS
a) EXAMINE IF OPEN
b) EXAMINE IF CLOSED
c) OUTPUT ENERGIZE

20a. _____

20b. _____

20c. _____

21. In general, a ladder logic rung consists of input conditions represented by (a) _____ symbol and an output instruction represented by the (b) _____ symbol.

21a. _____

21b. _____

22. When logic _____ exists in at least one path, the rung condition is said to be TRUE.

22. _____

23. The addressing format for inputs and outputs is standard for all PLC models. (True or False)

23. _____

24. Each contact instruction can be used only once throughout the program. (True or False)

24. _____

25. The _____ will indicate what PLC input is connected to what input device and what PLC output will drive what output device.

25. _____

26. There may be a limit to the number of series contact instructions that can be included in one rung of a ladder logic program. (True or False)

26. _____

27. With all PLCs, the only limit to the number of parallel branches per ladder logic rung is memory size. (True or False)

27. _____

28. An internal output does not directly control a(n) _____ device.

28. _____

29. The user program is not an electric circuit but a(n) _____ circuit.

29. _____

30. Both normally open and normally closed pushbuttons can be represented by the EXAMINE IF CLOSED instruction. (True or False)

30. _____

31. Normally, each PLC instruction requires one word of user memory. (True or False)

31. _____

32. Match the following typical PLC modes of operation with the most correct description. Place the number from the mode list in the answer column.

MODE
1) PROGRAM
2) TEST
3) RUN

DESCRIPTION
a) Used to execute the user program
b) Used to monitor the user program without energizing any outputs
c) Used to enter the user program

32a. _____

32b. _____

32c. _____

33. Match the following typical data files with the closest description. Place the number from the data file list in the answer column.

DATA FILES
1) Bit
2) Integer
3) Input
4) Status
5) Timer
6) Control

DESCRIPTION
a) Used for internal relay logic storage
b) Used for storage of the status of field devices
c) Used for the storage of accumulated and preset values
d) Stores controller operation information
e) Used to store numeric values
f) Used to store the length and pointer position for specific instructions

33a. _____

33b. _____

33c. _____

33d. _____

33e. _____

33f. _____

34. It takes the processor exactly the same amount of time to examine different types of instructions. (True or False)

34. _____

35. The time to scan the user program depends on the clock frequency of the microprocessor. (True or False)

35. _____

36. Sequential function chart programming allows the PLC ladder logic program to be structured into _____.

36. _____

37. On some PLC models, branches can be established at both input and output portions of a rung. (True or False)

37. _____

38. A(n) _____ branch starts or ends within another branch.

38. _____

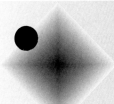

PROGRAMMING ASSIGNMENTS FOR CHAPTER 5

1. With reference to the programming manual for the PLC you will be working with, determine each of the following:
   a) Instructional address numbers used for external inputs
   b) Instructional address numbers used for external outputs
   c) Instructional address numbers used for internal control relays
   d) User word memory size
   e) Number of words required per instruction
   f) Typical scan time
   g) If nested branch circuits are permitted
   h) Maximum number of contacts that can be included in one rung of a ladder logic program
   i) Maximum number of parallel branches allowed per rung
   j) Number of outputs allowed per rung
   k) Location of the output on a rung

2. With reference to the programming manual for the PLC you will be working with, outline the key sequence required to enter each of the following program modes:
   a) PROGRAM
   b) TEST
   c) RUN

3. For each of the ladder logic programs shown in Fig. 5-12, do each of the following tasks:
   a) On a separate sheet of paper, redraw the ladder logic program, indicating the instructional address for all input and output devices (use normally open pushbuttons or switches for the input conditions and a pilot lamp for the output)

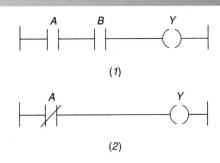

(1)

(2)

(3)

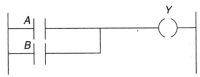

(4)

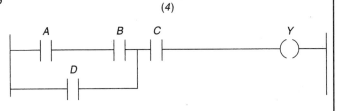

(5)

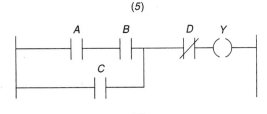

(6)

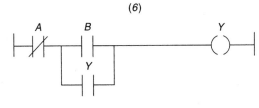

(7)

Fig. 5-12

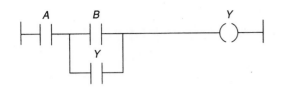

(8)

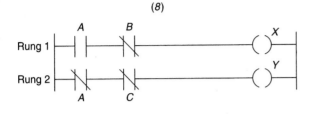

(9)

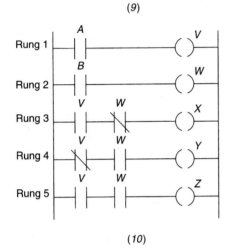

(10)

Fig. 5-12 (continued)

b) On a separate sheet of paper, draw the I/O wiring diagram for the circuit and wire the field devices to the PLC
c) Program the circuit into the controller
d) Operate the circuit in both the TEST and RUN modes and have both operations checked.
e) List what combination(s) of input(s) will produce an output at *Y*

4-1. On a separate sheet of paper, redraw the ladder logic circuit shown in Fig. 5-13 and program it to maintain the original control logic and eliminate the nested contact.

4-2. Program and wire the circuit into the PLC and have your instructor check it for operation. (Use normally open pushbuttons for the inputs and a pilot lamp for the output.)

4-3. List what combinations of inputs will produce an output at *Y.*

5-1. Assume that the PLC used to program the circuit in Fig. 5-14 can accommodate a maximum of five series contact instructions per rung. On a separate sheet of paper, redraw and program the circuit to meet this PLC requirement by making use of the internal relay instruction.

5-2. Program and wire the circuit into the PLC and have the instructor check it for operation. (Use normally open pushbuttons for inputs and a single pilot lamp for the output.)

6-1. What problem is posed in trying to program the circuit in Fig. 5-15 into some PLCs?

6-2. On a separate sheet of paper, redraw the program showing this problem corrected.

7-1. Explain the problem a PLC would have in attempting to execute the logic input instructions of the circuit in Fig. 5-16.

7-2. On a separate sheet of paper, redraw and program the circuit to avoid this problem.

7-3. Program and wire the circuit into the PLC and have your instructor check it for operation. (Use normally open pushbuttons for inputs and a pilot lamp for the output.)

7-4. List what combination of inputs will produce an output at *Y.*

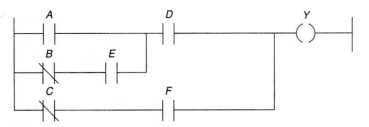

Fig. 5-13

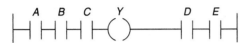

Fig. 5-14

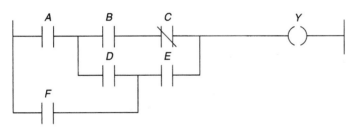

Fig. 5-15

Fig. 5-16

6

DEVELOPING FUNDAMENTAL PLC WIRING DIAGRAMS AND LADDER LOGIC PROGRAMS

TEST 6•1

Choose the letter that best completes the statement.

Answer

1. An electromagnet control relay is basically a(n):
   a) electromagnet used to switch contacts.
   b) electromagnet used to relay information.
   c) manually operated control device.
   d) pressure-operated control device.

1. _A_

2-1. In the relay illustration of Fig. 6-1, the coil would be considered:
   a) energized.
   b) de-energized.
   c) operated from an ac source.
   d) both *a* and *c*.

2-1. _A_

2-2. In the relay illustration of Fig. 6-1, contact No. 1 is an:
   a) NO fixed contact.
   b) NC fixed contact.
   c) NO movable contact.
   d) NC movable contact.

2-2. _B_

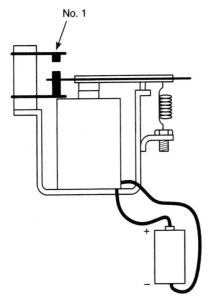

No. 1

Fig. 6-1

3. In the relay control circuit of Fig. 6-2, when the switch is *closed,* CR1 is:

3. C

   a) energized, and the red and green lights are both on.
   b) de-energized, the red light is off, and the green light is on.
   c) energized, the red light is on, and the green light is off.
   d) energized, the red light is off, and the green light is on.

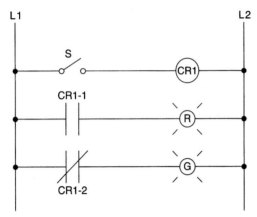

Fig. 6-2

4-1. In the motor starter circuit of Fig. 6-3, the main contacts M are:

4-1. D

   a) part of the power circuit.
   b) part of the control circuit.
   c) designed to handle the full load current of the motor.
   d) both *a* and *c.*

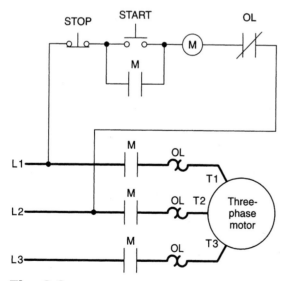

Fig. 6-3

4-2. In the motor starter circuit of Fig. 6-3, starter coil M is:

4-2. B

   a) part of the power circuit.
   b) energized to start the motor.
   c) energized only as long as the START button is pressed.
   d) all of the above.

4-3. In the motor starter circuit of Fig. 6-3, any overload current is sensed by the:
  a) starter coil.
  b) control contact M.
  c) OL coils.
  d) OL contact.

4-3. _D_

5. The abbreviations NO (normally open) and NC (normally closed) represent the electrical state of switch contacts when:
  a) power is applied.
  b) power is not applied.
  c) the switch is actuated.
  d) the switch is not actuated.

5. _B_

6. The pushbutton shown in Fig. 6-4 would be classified as a(n):
  a) NO pushbutton.
  b) NC pushbutton.
  c) break-before make pushbutton.
  d) ON/OFF pushbutton.

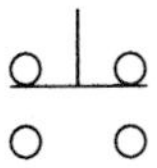

6. _C_

Fig. 6-4

7. The device represented by the symbol of Fig. 6-5 is a:
  a) drum switch.
  b) selector switch.
  c) sequence switch.
  d) toggle switch.

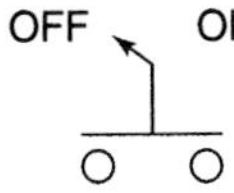

7. _B_

Fig. 6-5

8. A limit switch is usually actuated by:
  a) hand.
  b) pressure.
  c) contact with an object.
  d) an electromagnet.

8. _C_

9. A proximity switch can be actuated:
  a) without any physical contact.
  b) by a change in light intensity.
  c) by a change in capacitance.
  d) by all of the above.

9. _A_

10. Figure 6-6 represents the symbol for a:
  a) pressure switch.
  b) temperature switch.
  c) limit switch.
  d) level switch.

10. _C_

Fig. 6-6

11. Figure 6-7 represents the symbol for a:
    a) pressure switch.
    b) temperature switch.
    c) proximity switch.
    d) level switch.

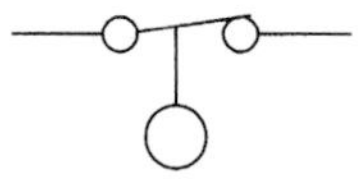

11. _D_

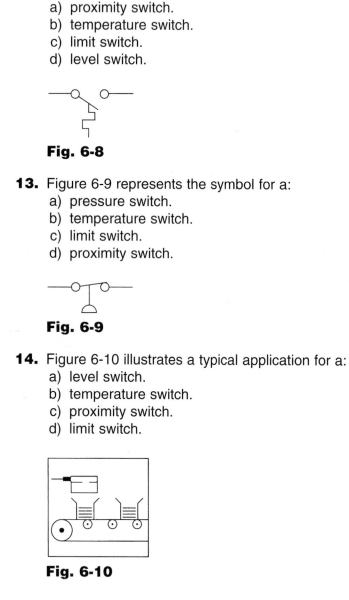

Fig. 6-7

12. Figure 6-8 represents the symbol for a:
    a) proximity switch.
    b) temperature switch.
    c) limit switch.
    d) level switch.

12. _B_

Fig. 6-8

13. Figure 6-9 represents the symbol for a:
    a) pressure switch.
    b) temperature switch.
    c) limit switch.
    d) proximity switch.

13. _A_

Fig. 6-9

14. Figure 6-10 illustrates a typical application for a:
    a) level switch.
    b) temperature switch.
    c) proximity switch.
    d) limit switch.

14. _C_

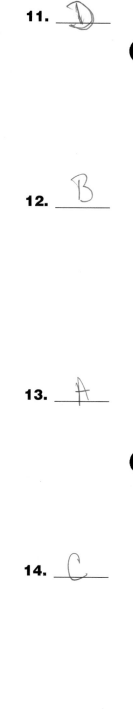

Fig. 6-10

15. Figure 6-11 represents the symbol for a:
a) heater.
c) solenoid.
b) horn.
d) motor.

Fig. 6-11

16. Figure 6-12 represents the symbol for a(n):
a) solenoid value.
c) overload relay coil.
b) motor starter.
d) overload relay contact.

Fig. 6-12

17. The circuit of Fig. 6-13 is that of an electromagnetic:
a) control relay.
c) motor starter relay.
b) overload relay.
d) latching relay.

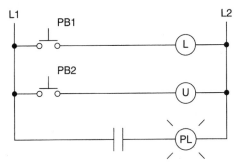

Fig. 6-13

18. In the circuit of Fig. 6-13, the pilot light is switched on by:
a) continually pressing PB1.
c) momentarily pressing PB1.
b) continually pressing PB2.
d) either *a* or *c*.

18. _____

19. The correct ladder logic program for the circuit in Fig. 6-13 is shown in which of the following?

19. _____

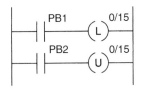

(a)

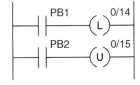

(c)

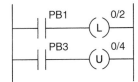

(b)   *(d)*

20-1. In the I/O wiring diagram and ladder logic program of Fig. 6-14, what is the three-digit address number of the instruction associated with the pressure switch?

20-1. _____

    a) I/0   c) I/10

    b) I/3   d) I/2

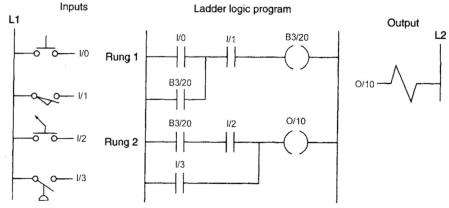

Fig. 6-14

20-2. In Fig. 6-14, what is the type of the instruction associated with the selector switch?

20-2. _____

    a) OUTPUT ENERGIZE   c) EXAMINE IF OPEN

    b) OUTPUT DE-ENERGIZE   d) EXAMINE IF CLOSED

20-3. In Fig. 6-14, rung 1 will be TRUE when:

20-3. _____

    a) the pushbutton and limit switch are actuated.

    b) the selector switch is turned ON and the pressure switch is actuated.

    c) the selector switch is turned ON and the limit switch is actuated.

    d) either *a* or *c*.

20-4. In Fig. 6-14, rung 2 will be TRUE when:

20-4. _____

    a) the pressure switch is actuated.

    b) the pushbutton is actuated.

    c) the selector switch is turned ON and rung 1 is TRUE.

    d) either *a* or *c*.

20-5. In Fig. 6-14, the instruction at address B3/20 is associated with:

20-5. _____

    a) an internal relay coil.   c) an external output device.

    b) an external input device.   d) the solenoid.

20-6. In Fig. 6-14, if the EXAMINE IF CLOSED instruction at address I/3 is TRUE:

20-6. _____

    a) output B3/20 will also be TRUE.

    b) output O/10 will also be TRUE.

    c) input I/2 will also be TRUE.

    d) both *a* and *b*.

20-7. In Fig. 6-14, assume that an NC limit switch was substituted for the one shown. For the circuit to operate in the same manner as before:
   a) the wires to the limit switch would have to be reversed.
   b) the address number would have to be changed to I/4.
   c) the instruction representing it would have to be changed to an EXAMINE IF OPEN.
   d) the instruction representing it would have to be changed to an OUTPUT ENERGIZE.

20-7. _____

21. In Fig. 6-15, when LS2 is actuated and LS1 is not:
   a) input 002 is TRUE.
   b) user program rung 2 is FALSE.
   c) pilot light PL2 is ON.
   d) both *a* and *c*.

21. _____

22. In Fig. 6-15, when LS1 is actuated and LS2 is not:
   a) pilot light PL1 is ON.
   b) pilot light PL2 is ON.
   c) pilot light PL3 is ON.
   d) all of the above.

22. _____

23. In Fig. 6-15, when both LS1 and LS2 are not actuated:
   a) user program rung 1 is FALSE.
   b) user program rung 2 is TRUE.
   c) pilot light PL2 is ON.
   d) all of the above.

23. _____

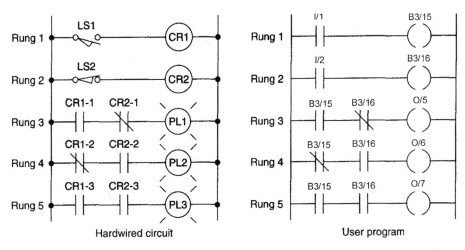

Hardwired circuit   User program

Note: Limit switches LS1 and LS2, and pilot lights PL1, PL2, and PL3 are represented in the user program by addresses I/1, I/2, O/5, O/6, and O/7, respectively.

Fig. 6-15

24. In Fig. 6-15, when both LS1 and LS2 are actuated:
   a) user program rung 2 is TRUE.
   b) user program rung 3 is TRUE.
   c) user program rung 4 is TRUE.
   d) user program rung 5 is TRUE.

24. _____

25-1. In the ladder logic program of Fig. 6-16, output O/10 is energized when:
    a) input I/1 is TRUE.
    b) input I/2 is TRUE.
    c) input I/3 is TRUE.
    d) inputs I/1 and I/2 are TRUE.

25-1. _____

25-2. In Fig. 6-16, assume that rung 2 is FALSE and rung 1 makes a transition from TRUE to FALSE. As a result, output O/10:
    a) remains ON.
    b) remains OFF.
    c) switches from ON to OFF.
    d) switches from OFF to ON.

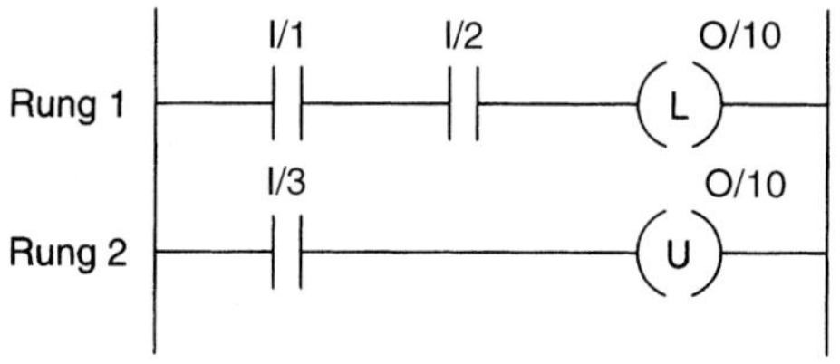

25-2. _____

Fig. 6-16

25-3. The circuit in Fig. 6-16 is retentive on power lost. This means that if power is lost, then output at address O/10 will:
    a) switch to the ON state.
    b) switch to the OFF state.
    c) remain in its original ON or OFF state.
    d) flash ON and OFF.

25-3. _____

25-4. In Fig. 6-16, if inputs I/1, I/2, and I/3 are all TRUE, the output at address O/10 will:
    a) switch to the ON state.
    b) switch to the OFF state.
    c) remain in its original ON or OFF state.
    d) flash ON and OFF.

25-4. _____

26. A contractor
    a) is another name for a relay.
    b) is designed to handle heavy power loads.
    c) always has an overload relay physically and electrically attached.
    d) is a physically small relay.

26. _____

27. An inductive proximity sensor is actuated by:
    a) a metal object.   c) a light beam.
    b) a nonconductive material.   d) all of the above.

27. _____

28. The most common actuator for a reed switch is:
    a) a permanent magnet.
    b) a light beam.
    c) application of pressure.
    d) a heat source.

28. _____

29. A(n) _____ converts light energy directly into electric energy.   29. _____
    a) LED
    b) phototransistor
    c) solar cell
    d) photoconductive cell

30. The light source used in most industrial photoelectric sensors is a(n):   30. _____
    a) LED.
    b) phototransistor.
    c) photovoltaic cell.
    d) miniature incandescent lamp.

31. A(n) _____ operates by sending sound waves toward the target   31. _____
and measuring the time it takes for the pulses to bounce back.
    a) pressure sensor
    b) bar code scanner
    c) ultrasonic sensor
    d) flowmeter

32. The force applied to a strain gauge causes it to bend and change its: 32. _____
    a) temperature.
    b) resistance.
    c) voltage.
    d) current.

33. A thermocouple, when heated: 33. _____
    a) produces a small dc voltage.
    b) produces a small ac voltage.
    c) increases its resistance value.
    d) decreases its resistance value.

34. A tachometer normally refers to a(n) _____ used for speed   34. _____
measurement.
    a) load cell
    b) capacitive proximity sensor
    c) ultrasonic sensor
    d) dc generator

35. Solenoid valves are available to control: 35. _____
    a) oil flow.
    b) air flow.
    c) water flow.
    d) all of the above.

36. A(n) _____ converts electrical pulses applied to it into discrete   36. _____
rotor movements.
    a) tachometer
    b) solenoid
    c) stepper motor
    d) electronic magnetic flowmeter

TEST 6 • 2

Place the answers to the following questions in the answer column at the right.

Answer

1. An electrical relay uses _____ to switch contacts.

1. _____

2. When current flows through the coil of a relay, the coil is said to be _____.

2. _____

3. A normally closed (NC) relay contact is closed when current flows through the coil.
(True or False)

3. _____

4. A relay usually will have only one coil but a number of different contacts. (True or False)

4. _____

5. Each contact of a relay is usually drawn as it would appear with the coil _____.

5. _____

6. In a magnetically operated, across-the-line ac starter, the control circuit is required to handle the full load current of the motor. (True or False)

6. _____

7. In a motor starter, a(n) _____ relay is provided to protect the motor against current overloads.

7. _____

8. A(n) _____ operated switch is controlled by hand.

8. _____

9. A limit switch is classified as a(n) _____ operated switch.

9. _____

10. Match the symbols in Fig. 6-17 with the correct switch designation. Place the number from the symbols list in the answer column.

SYMBOLS

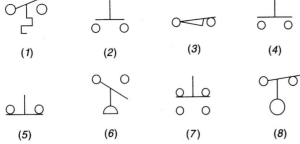

(1)   (2)   (3)   (4)

(5)   (6)   (7)   (8)

Fig. 6-17

SWITCH DESIGNATIONS
a) NO pushbutton
b) NC pushbutton
c) Break-make pushbutton
d) Selector switch
e) NC limit switch
f) NO temperature switch
g) NO pressure switch
h) NC level switch

10a. _____

10b. _____

10c. _____

10d. _____

10e. _____

10f. _____

10g. _____

10h. _____

11. Proximity switches usually sense the presence or absence of a target by means of physical contact. (True or False)

11. _____

12. Match the symbols in Fig. 6-18 with the correct output device. Place the number from the symbols list in the answer column.

SYMBOLS

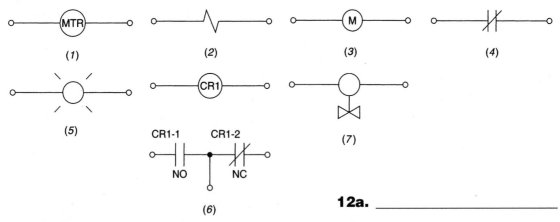

(1)   (2)   (3)   (4)

(5)

CR1-1   CR1-2
NO   NC

(6)

(7)

Fig. 6-18

12a. _____

12b. _____

12c. _____

OUTPUT DEVICE
  a) Pilot light
  b) Control relay
  c) Motor starter
  d) OL relay contact
  e) Solenoid
  f) Solenoid valve
  g) Motor

12d. _____

12e. _____

12f. _____

12g. _____

13. Latching relays are used when it is necessary for contacts to stay open and/or closed, even though the coil is momentarily energized. (True or False)

13. _____

14. The electromagnetic latching relay function can be programmed on a PLC using the (a) _____ and (b) _____ coil instruction.

14a. _____

14b. _____

15. The programmed latching relay instruction is retentive; that is, if the relay is latched, it will unlatch if power is lost and then restored. (True or False)

15. _____

16. _____ controls are required for processes that demand certain operations be performed in a specific order.

16. _____

17. _____ controls require that certain operations be performed without regard to the order in which they are performed.

17. _____

18. Control relays are designed to handle heavy currents and high voltages. (True or False)

18. _____

19. Programmable controllers have I/O capable of operating heavy power loads directly. (True or False)

19. _____

20. The magnetic motor starter is a contactor with a(n) _____ physically and electrically attached.

20. _____

21. Changes to dual in-line (DIP) package switch settings occur mainly during installation. (True or False)

21. _____

22. The difference between the ON and OFF point of a proximity sensor is known as _____.

22. _____

23. A small current flows through a solid-state proximity sensor even when the output is turned OFF. (True or False)

23. _____

24. An inductive proximity sensor is actuated by conductive and nonconductive materials. (True or False)

24. _____

25. LED light sources used in photoelectric sensors are normally ON continually. (True or False)

25. _____

26. A through-beam photoelectric sensor is used to detect the light beam reflected for a target. (True or False)

26. _____

27. Bar-code _____ are available for PLCs to read the bar-codes on boxes.

27. _____

28. For a thermocouple to generate a voltage, a temperature difference must exist between the (a) _____ and (b) _____ junctions.

28a. _____

28b. _____

29. The usual approach to flow measurement is to convert the _____ energy that the fluid has into some other measurable form.

29. _____

30. A magnetic pickup sensor generates voltage pulses, and the _____ of these pulses is used to determine speed.

30. _____

31. A solenoid is made up of a (a) _____ with a (b) _____ iron core.

31a. _____

31b. _____

32. A one-degree-per-step stepper motor requires _____ pulses to move through one revolution.

32. _____

33. A _____ circuit is a method of maintaining current flow after a momentary switch has been pressed and released.

33. _____

34. There is more than one correct way to implement the ladder logic for a given control process. (True or False)

34. _____

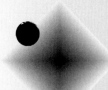

PROGRAMMING ASSIGNMENTS
FOR CHAPTER 6

This section will present several common programming conversion applications designed to give you, the student, a feel for the potential of the ladder logic programming language. The instructions used are intended to be generic in nature and, as such, will require some conversion for the particular PLC model you are using. The use of a prewired PLC input/output control panel is recommended to simulate the operation of these circuits.

1. The circuit in Fig. 6-19 demonstrates that the contacts of a programmed output relay can be examined for an ON or OFF condition as many times as you like. Prepare an I/O connection diagram and ladder logic program for the circuit. Enter the program into the PLC and prove its operation.

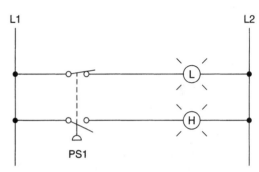

Fig. 6-20

3. The circuit in Fig. 6-21 illustrates how a standard START/STOP motor control circuit can be programmed using a PLC. Prepare an I/O connection diagram and ladder logic program for the circuit. Enter the program into the PLC and prove its operation.

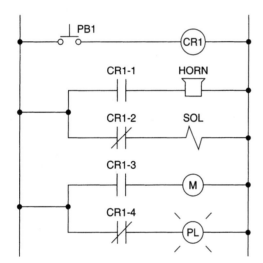

Fig. 6-19

2. The circuit in Fig. 6-20 demonstrates that the contacts of a single-pole input device can be programmed as a double-pole input device. Prepare an I/O connection diagram and ladder logic program for the circuit using only the *single* set of NO contacts of the pressure switch.

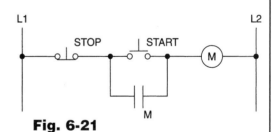

Fig. 6-21

4. The circuit in Fig. 6-22 illustrates how multiple starts and stops are programmed. Prepare an I/O connection diagram and ladder logic program for the circuit. Enter the program into the PLC and prove its operation.

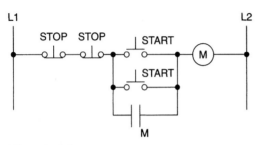

Fig. 6-22

5. The circuit in Fig. 6-23 illustrates how a manual/automatic control circuit can be programmed using a PLC. The pump is started by pressing the start button. With the selector switch in the manual position, the solenoid valve is energized at *all* times. With the selector switch in the automatic position, the solenoid valve is energized only when the pressure switch is closed. Prepare an I/O connection diagram and ladder logic program for the circuit. Enter the program into the PLC and prove its operation.

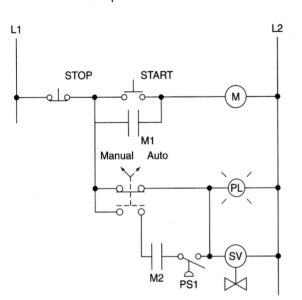

Fig. 6-23

6. The circuit in Fig. 6-24 illustrates how a forward/reverse motor starter, with electrical interlocks, can be programmed using a PLC. Prepare an I/O connection diagram and ladder logic program for the circuit. Enter the program into the PLC and prove its operation.

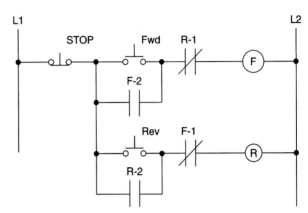

Fig. 6-24

7. The circuit in Fig. 6-25 illustrates how an electric door opener, with top and bottom travel limit switches, can be programmed using a PLC. Prepare an I/O connection diagram and ladder logic program for the circuit. Enter the program into the PLC and prove its operation.

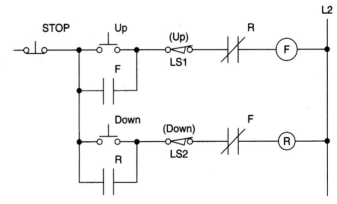

Fig. 6-25

8. The circuit in Fig. 6-26 illustrates how a reciprocating motion machine process can be programmed using a PLC. The workpiece starts on the left and moves to the right when the START button is pressed. When it reaches the rightmost limit, the drive motor reverses and brings the workpiece back to the leftmost position again, and the process repeats. The forward and reverse pushbuttons provide a means of starting the motor in either forward or reverse so that the limit switches can take over automatic control. Prepare an I/O connection diagram and ladder logic program for the circuit. Enter the program into the PLC and prove its operation.

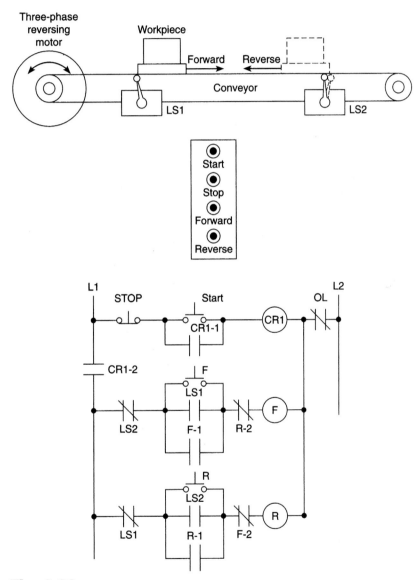

Fig. 6-26

9. The circuit in Fig. 6-27 illustrates how the sequence of operation of two motors can be electrically controlled by programming of the PLC. This circuit requires that motor No. 2 *cannot* be started unless motor No. 1 is running. Prepare an I/O connection diagram and ladder logic program for the circuit. Enter the program into the PLC and prove its operation.

10. The schematic in Fig. 6-28 illustrates the use of the EXCLUSIVE-OR circuit applied to a pushbutton interlock circuit. The output is energized if button A or button B is pressed but *not* if *both* are pressed. Prepare an I/O connection diagram and ladder logic program for the circuit. Enter the program into the PLC and prove its operation.

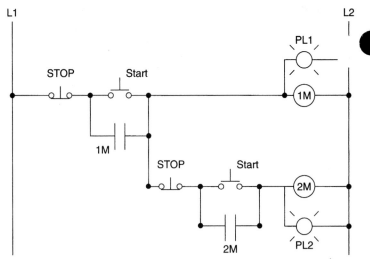

Fig. 6-27

11-1. The ladder logic program in Fig. 6-29 is a push-to-start/push-to-stop circuit. A single pushbutton performs both the START and STOP functions. The first time you press the pushbutton (represented by address I:0/1), instruction B3/11 is latched, energizing output O:0/2. The second time you press the pushbutton, instruction B3/12 unlatches instruction B3/11, de-energizing output O:0/2. Instruction B3/10 prevents interaction between instructions B3/12 and B3/11. Prepare an I/O connection diagram and ladder logic program for the circuit. Use addresses that apply to your PLC. Enter the program into the PLC and prove its operation.

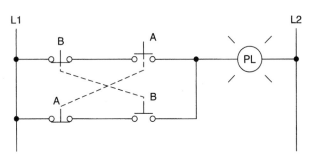

Fig. 6-28

Ladder logic program

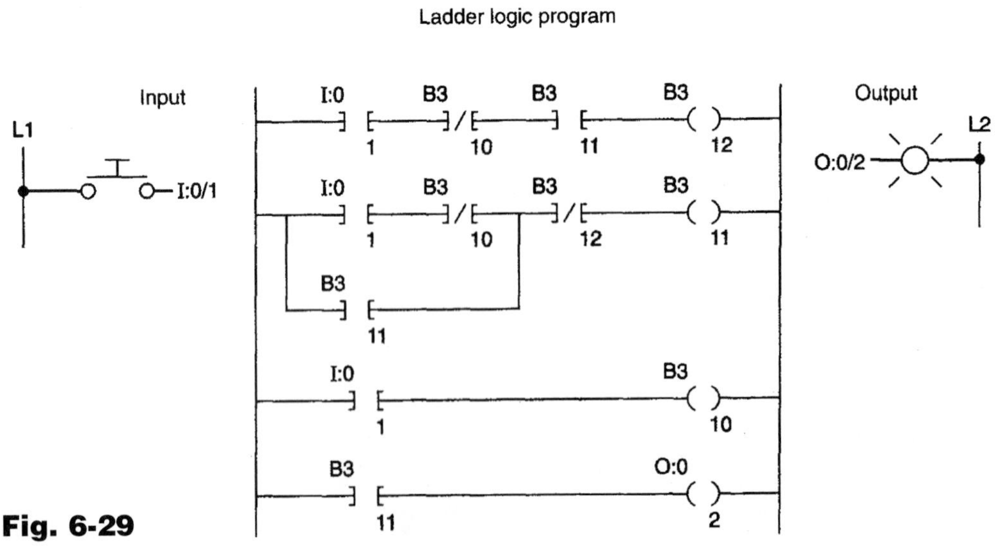

Fig. 6-29

11-2. The program in Fig. 6-29 is to be used to control the light ON and OFF from four remote locations. Assuming that one NO pushbutton is used at each location, prepare an I/O connection diagram and ladder logic program for the circuit. Enter the program into the PLC and prove its operation.

12-1. Prepare an I/O connection diagram and a ladder logic program for the LATCH/UNLATCH program shown in Fig. 6-30. Enter the program into the PLC and prove its operation.

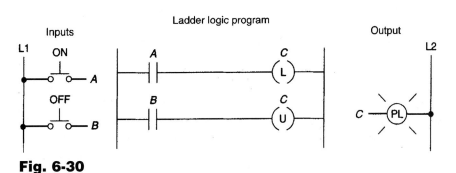

Fig. 6-30

12-2. Operate the circuit with *both* the ON (LATCH) and OFF (UNLATCH) pushbutton pressed. Make note of the status of the light (ON or OFF at all times?). With reference to the way the controller executes the program, explain why the light appears to be ON or OFF at all times.

12-3. Repeat 12-2 with the order of the program changed as shown in the ladder logic program in Fig. 6-31.

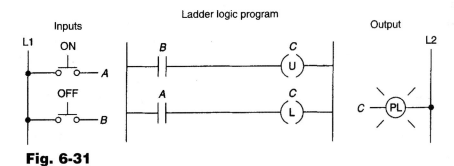

Fig. 6-31

13. Design two start/stop motor control circuit programs that can achieve the desired result of Fig. 6-32. One design is to be a seal-in type, start/stop control, and the other is to be a latch/unlatch type start/stop control. Use one program to accomplish the task. Enter the program into the PLC and prove its operation. With both outputs initially ON, simulate a power failure to the

PLC system. When power was restored, what happened to the outputs? When would you use one circuit over the other?

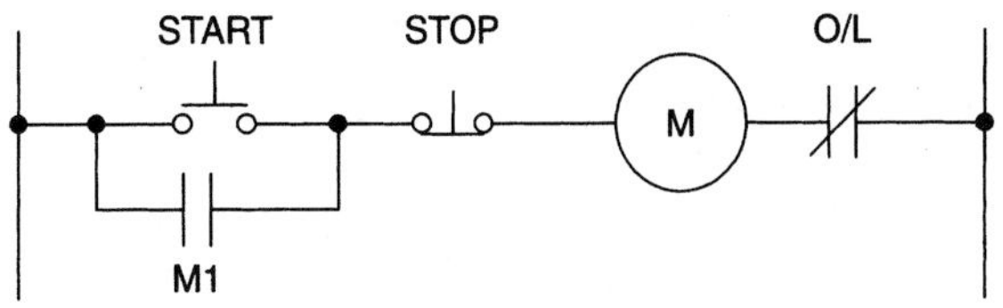

```
      START      STOP            O/L
  ●────o o────●──o o────( M )────|/|────●
       │      │
       ├─┤ ├──┘
       M1
```

Fig. 6-32

14. Design a program that will perform the following tasks:

    a) When one switch is closed, three lights will come on.

    b) If another switch is closed, two of the lights will drop out, leaving one light on.

    c) When another switch is selected (all three switches on at this point), one of the lights that dropped out from condition *b* above will come on, thus showing two lights on.

    d) A fourth switch turns off any of the lights that happen to be on.

Create two programs to achieve the desired result. Using the memory map feature of the PLC software package, compare the number of words of memory used by each arrangement.

15. There are four input sensors to an annunciator system that switches the output ON if some operational malfunction occurs. Design a program that operates the alarm system as follows:

    • If any one input is ON, nothing happens.

    • If any two inputs are ON, a green pilot light goes ON.

    • If any three inputs are ON, a yellow pilot light goes ON.

    • If all four inputs are ON, a red pilot light goes ON.

16. Prepare an I/O connection diagram and a ladder logic program for the process control circuit described in Fig. 6-52 of the text. Enter the program into the PLC and prove its operation.

17. Prepare an I/O connection diagram and a ladder logic program for the process control circuit described in Fig. 6-55 of the text. Enter the program into the PLC and prove its operation.

18. Prepare an I/O connection diagram and a ladder logic program for the process control circuit described in Fig. 6-58 of the text. Enter the program into the PLC and prove its operation.

19. Prepare an I/O connection diagram and a ladder logic program for the process control circuit described in Fig. 6-59 of the text. Enter the program into the PLC and prove its operation.

20. Prepare an I/O connection diagram and a ladder logic program for the process control circuit described in Fig. 6-60 of the text. Enter the program into the PLC and prove its operation.

21. Write a ladder logic program that will cause output pilot light PL to be ON when selector switch SS is closed, pushbutton PB is open, and limit switch LS is open. Enter the program into the PLC and prove its operation.

22. Write a ladder logic program that will cause a solenoid, SOL, to be energized when limit switch LS is closed and pressure switch PS is open. Enter the program into the PLC and prove its operation.

23. Write a ladder logic program that will cause output pilot light PL to be latched when pushbutton PB1 is closed, and unlatched when either pushbutton PB2 or pushbutton PB3 is closed. Also, do not allow the unlatch to go TRUE when the latch rung is TRUE, nor allow the latch rung to go TRUE when the unlatch rung is TRUE. Enter the program into the PLC and prove its operation.

7 PROGRAMMING TIMERS

TEST 7•1

Choose the letter that best completes the statement.   Answer

1. Certain contacts of a mechanical timing relay are designed to operate   **1.** _____
   at a preset time interval:
   a) after the coil is energized.
   b) after the coil is de-energized.
   c) after power is applied to the circuit.
   d) either *a* or *b.*

2. Which of the following symbols represents an on-delay timed   **2.** _____
   relay contact?

   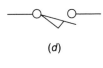

    (*a*)   (*b*)   (*c*)   (*d*)

3. The relay contact drawn in Fig. 7-1 is designed to operate so that: **3.** _____
   a) when the relay coil is energized, there is a time delay in closing.
   b) when the relay coil is energized, there is a time delay in opening.
   c) when the relay coil is de-energized, there is a time delay before the
      contact opens.
   d) when the relay coil is de-energized, there is a time delay before the
      contact closes.

Fig. 7-1

4. In the circuit in Fig. 7-2, the light will stay on: **4.** _____
   a) as long as S1 is closed.
   b) for 5 s after coil TD is energized.
   c) for 5 s after coil TD is de-energized.
   d) both *a* and *c.*

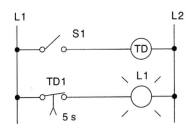

Fig. 7-2

5-1. The timer instruction used in the PLC program of Fig. 7-3 is the
_____ type.
a) block-formatted
b) pneumatic
c) coil-formatted
d) motor-driven

Rung 1

```
      001      002                901
   ──┤ ├──────┤ ├─────────────(TON)──
                              PR: 0500
                              TB: 0.1 s
                              AC: 000
```

Rung 2

```
      901                        009
   ──┤ ├──────────────────────( )──
```

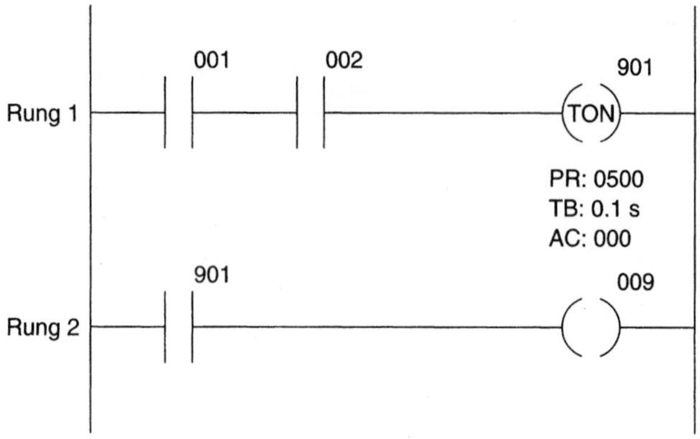

Fig. 7-3

5-2. The timer wilt start timing whenever:
a) input 001 is on.
b) input 002 is on.
c) both input 001 and 002 are on.
d) power is applied to the circuit.

5-2. _C_

5-3. The timer will reset whenever:
a) input 001 is off.
b) input 002 is off.
c) power is lost.
d) all of the above.

5-3. _D_

5-4. The time delay period for this timer is:
a) 500 s.
b) 500 min.
c) 50 s.
d) 50 min.

5-4. _C_

5-5. In this program, output 009 is energized whenever:
a) rung No. 1 is true.
b) rung No. 2 is false.
c) the accumulated time equals the preset time.
d) all of the above.

5-5. _C_

6-1. According to the relay schematic diagram in Fig. 7-4, coil TD1
is to be energized:
a) the instant the start button is pressed.
b) the instant the stop button is pressed.
c) 5 s after the start button is pressed.
d) 5 s after the stop button is pressed.

6-1. _A_

6-2. According to the relay schematic diagram in Fig. 7-4, contact TD1-2:   **6-2.** _D_
  a) opens 5 s after coil TD1 energizes.
  b) closes 5 s after coil TD1 de-energizes.
  c) opens 5 s after coil TD1 de-energizes.
  d) closes 5 s after coil TD1 energizes.

6-3. According to the ladder logic program in Fig. 7-4, the address of the   **6-3.** _B_
start pushbutton would be:
  a) 001.
  b) 002.
  c) 009.
  d) 033.

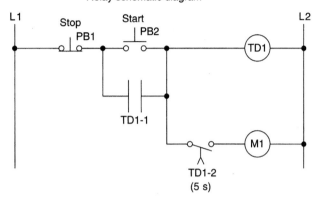

Relay schematic diagram

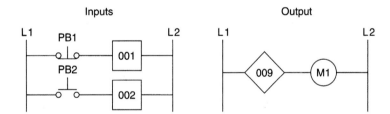

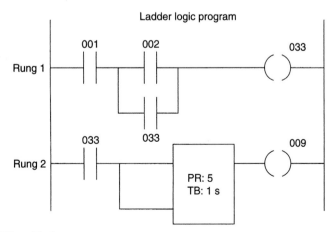

Ladder logic program

Fig. 7-4

6-4. According to the ladder logic program in Fig. 7-4, the stop pushbutton would have a(n):
    a) normally open contact.
    b) normally closed contact.
    c) break before make contact.
    d) either *a* or *b*.

6-4.

6-5. According to the ladder logic program in Fig. 7-4, the relay addressed 033 is an:
    a) external relay used to operate coil M1.
    b) external relay used to control the timer.
    c) internal relay used to operate coil M1.
    d) internal relay programmed to replace the function of relay contact TD1-1.

6-5. D

6-6. According to the ladder logic program in Fig. 7-4, once the timer has timed out:
    a) rung 1 will be true.
    b) rung 2 will be true.
    c) output 009 will remain energized.
    d) all of the above.

6-6. d

6-7. According to the ladder logic program in Fig. 7-4, the timer is reset anytime:
    a) input 001 is off.
    b) input 002 if off.
    c) input 003 is on.
    d) all of the above.

6-7. A

7. Which of the following ladder logic programs could be used to implement the relay schematic circuit of Fig. 7-5?

7. B C

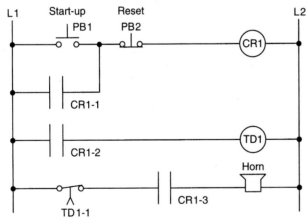

Fig. 7-5

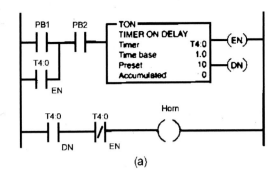

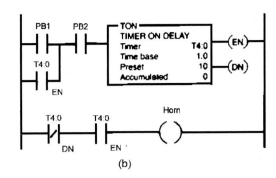

(a)   (b)

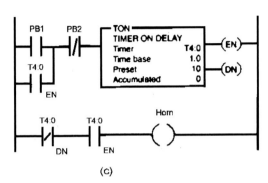

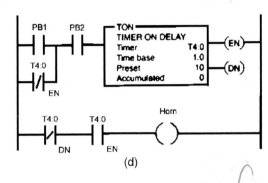

(c)   (d)

8-1. The type of timer programmed in Fig. 7-6 is a:

    a) retentive on-delay.
    b) retentive off-delay.
    c) nonretentive off-delay.
    d) nonretentive on-delay.

8-1. _____

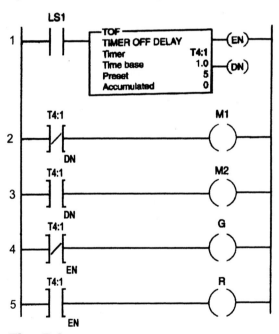

Fig. 7-6

8-2. In Fig. 7-6, the timer begins timing when:

    a) rung No. 1 makes a transition from false to true.
    b) rung No. 1 makes a transition from true to false.
    c) rung No. 4 makes a transition from false to true.
    d) rung No. 4 makes a transition from true to false.

8-2. _____

8-3. Rung No. 4 will be true when:
  a) input LS1 is on.
  b) input LS1 is off.
  c) rung 1 has been false for 5 s.
  d) rung 1 has been true for 5 s.

8-4. Rung No. 3 is true:
  a) after the accumulated time value reaches the preset time value.
  b) until the accumulated time value reaches the preset time value.
  c) when rung No. 3 is true.
  d) on power-up only.

8-5. Output R will be energized when:
  a) rung No. 1 is true.
  b) rung No. 2 is false.
  c) rung No. 3 is true.
  d) rung No. 4 is false.

8-6. Output M1 will be energized when:
  a) rung No. 3 is false.
  b) rung No. 2 is true.
  c) LS1 is on.
  d) both *a* and *b.*

9. The operation of a PLC retentive timer is similar to that of an:
  a) electromagnetic pneumatic timer.
  b) electromechanical motor-driven timer.
  c) off-delay timer.
  d) on-delay timer.

10. The main difference between a PLC retentive and nonretentive timer is that the:
  a) retentive timer can be programmed for much longer time delay periods.
  b) nonretentive time can be programmed for much longer time delay periods.
  c) retentive timer maintains the current time should power be removed from the device or when the timer rung goes false.
  d) nonretentive timer maintains the current time should power be removed from the device or when the timer rung goes false.

11. Unlike the TON timer, the RTO timer requires a(n):
  a) timer reset instruction.
  b) input condition instruction.
  c) internal relay instruction.
  d) instantaneous contact instruction.

12. When addressing an RES instruction, it must be addressed to:
  a) a TOF instruction.
  b) a TON instruction.
  c) any word other than the RTO instruction.
  d) the same word as the RTO instruction.

8-3. _____

8-4. _A___

8-5. _C___

8-6. _D___

9. _B___

10. _C___

11. _____

12. _____

13. The main difference between a TON and TOF timer is that the:

    a) TON timer can maintain its accumulated time on loss of power or logic continuity.

    b) TOF timer can maintain its accumulated time on loss of power or logic continuity.

    c) TOF timer begins timing when logic continuity to the timing rung is lost.

    d) TON timer begins timing when logic continuity to the timing rung is lost.

13. _____

14. The type of timer programmed in Fig. 7-7 is a:

    a) retentive on-delay.

    b) retentive off-delay.

    c) nonretentive off-delay

    d) nonretentive on-delay.

14. _____

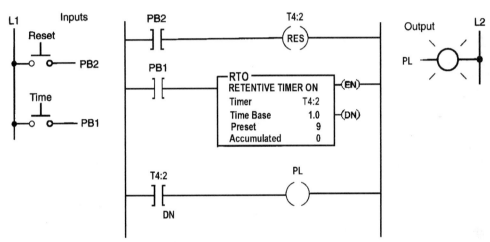

Fig. 7-7

15. The timer of Fig. 7-7 starts timing when:

    a) PB1 is closed.

    b) PB2 is closed.

    c) both PB1 and PB2 are closed.

    d) either PB1 or PB2 is closed.

15. _____

16. The timer of Fig. 7- 7 is reset when:

    a) PB1 is closed.

    b) PB2 is closed.

    c) PB1 is open.

    d) PB2 is open.

16. _____

17. In the timer program of Fig. 7-7, assume the following sequence of events:

    1) PB2 is momentarily pressed closed.

    2) PB1 is pressed closed for 5 s and released.

    3) PB2 is pressed closed for 4 s and released.

17-1. As a result, the timer-accumulated value at the end of the sequence would be:
   a) 005.
   b) 004.
   c) 009.
   d) 000.

17-1. _____

17-2. As a result, output PL would be:
   a) on for 4 s and off for 5 s.
   b) on for 5 s and off for 4 s.
   c) on after the entire sequence has been completed.
   d) off after the entire sequence has been completed.

17-2. _____

18. In the timer program of Fig. 7-7, assume the following sequence of events:
   1) PB2 is momentarily pressed closed.
   2) PB1 is pressed closed for 3 s and released.
   3) PB1 is again pressed closed for 6 s and released.

18-1. As a result, the timer-accumulated value at the end of the sequence would be:
   a) 003.
   b) 006.
   c) 009.
   d) 000.

18-1. _____

18-2. As a result, output PL would be:
   a) on for 3 s and off for 6 s.
   b) off for 3 s and on for 6 s.
   c) on after the entire sequence has been completed.
   d) off after the entire sequence has been completed.

18-2. _____

19. To reset a retentive timer, the:
   a) AC time must be greater than the PR time.
   b) PR time must be greater than the AC time.
   c) AC time must equal the PR time.
   d) none of the above.

19. _____

20. If the control input and reset input to a retentive timer both have logic continuity, the timer will:
   a) continue timing.
   b) stop timing but maintain its accumulated time value.
   c) stop timing and reset.
   d) continue timing until the PR time value equals the AC time value.

20. _____

21. The interconnecting of timers is commonly called:
   a) grouping.
   b) programming.
   c) sequencing.
   d) cascading.

21. _____

22. In the timer program of Fig. 7-8, the timer 901 is energized by actuating:
        a) PB1.
        b) PB2.
        c) both PB1 and PB2.
        d) either PB1 or PB2.

22. _____

Ladder logic program

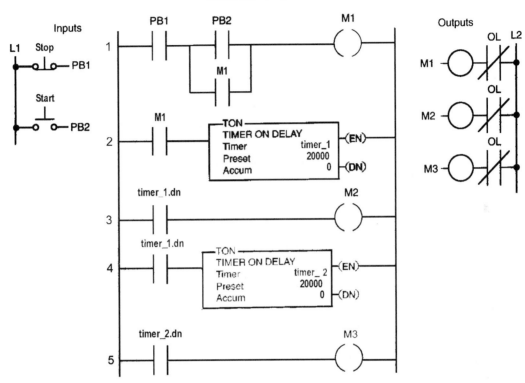

Fig. 7-8

23-1. In the timer program of Fig. 7-8, output M1 is normally energized:
        a) as soon as PB1 is actuated.
        b) as soon as PB2 is actuated.
        c) 10 s after PB1 has been actuated.
        d) 10 s after both PB1 and PB2 have been actuated.

23-1. _____

23-2. In Fig. 7-8, output M2 is normally energized _____ s after output M1 has been energized.
        a) 5
        b) 10
        c) 15
        d) 20

23-2. _____

23-3. In Fig. 7-8, output M3 is energized when:
        a) rungs 1, 2, 3 are true and rungs 4, 5 are false.
        b) rungs 1, 2, 3 are false and rungs 4, 5 are true.
        c) rungs 1, 2, 3, 4 are true and rung 5 is false.
        d) all rungs are true.

23-3. _____

24-1. In the program of Fig. 7-9, the two internal timers are interconnected to form a(n):
    a) amplifier circuit.
    b) rectifier circuit.
    c) oscillator circuit.
    d) series/parallel circuit.

24-1. _____

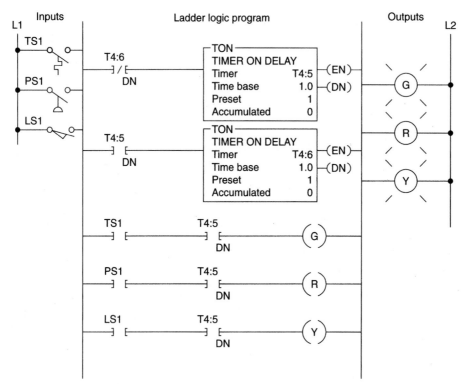

Fig. 7-9

24-2. The output of timer T4:5 in Fig. 7-9:
    a) turns on after a 015-s time delay and remains on.
    b) turns off after a 015-s time delay and remains off.
    c) turns on after a 0.1-s time delay and remains on.
    d) is pulsed on and off.

24-2. _____

24-3. In Fig. 7-9, when pressure switch PS1 closes, the:
    a) green indicating lamp is turned on.
    b) red indicating lamp is turned on.
    c) yellow indicating lamp is pulsed on and off.
    d) red indicating lamp is pulsed on and off.

24-3. _____

25-1. The timer program of Fig. 7-10 is designed to produce a:
    a) shorter time delay period than a single timer will allow.
    b) longer timer delay period than a single timer will allow.
    c) flasher light circuit.
    d) sequence of timed events.

25-1. _____

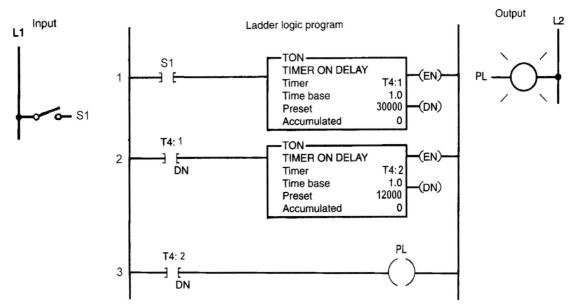

Fig. 7-10

25-2. Rung No. 1 is true whenever:
   a) switch S1 is closed.
   b) timer T4:2 is energized.
   c) output PL is on.
   d) all of the above.

25-3. Rung No. 2 will be true:
   a) whenever S1 is closed.
   b) 30,000 s after S1 is closed.
   c) 12,000 s after S1 is closed.
   d) 42,000 s after S1 is closed.

25-4. Rung No. 3 will be true:
   a) whenever S1 is closed.
   b) 30,000 s after S1 is closed.
   c) 12,000 s after S1 is closed.
   d) 42,000 s after S1 is closed.

25-2. _____

25-3. _____

25-4. _____

TEST 7•2

Place the answers to the following questions in the answer column at the right.

Answer

1. Mechanical timing relays are used to _____ the opening or closing of contacts for circuit control.

1. _____

2. An off-delay timing relay provides time delay when its coil is _____.

2. _____

3. Match each of the timed contact symbols with the correct contact operation. Place the number from the symbols list in the answer column.

SYMBOLS

   (1)   (2)   (3)   (4)

CONTACT OPERATION

a) When the relay is energized, there is a time delay before the contact closes.

3a. _____

b) When the relay is energized, there is a time delay before the contact opens.

3b. _____

c) When the relay is de-energized, there is a time delay before the contact opens.

3c. _____

d) When the relay coil is de-energized, there is a time delay before the contact closes.

3d. _____

4. PLC timers are input instructions that provide the same functions as mechanical timing relays. (True or False)

4. _____

5. Timer instructions are found on all PLCs manufactured today. (True or False)

5. _____

6. The number of timer instructions that can be programmed on a PLC is limited. (True or False)

6. _____

7. Timer instructions may be (a) _____ formatted or (b) _____ formatted.

7a. _____

7b. _____

8. Usually included as part of a timer instruction are the timer (a) _____, (b) _____, (c) _____, (d) _____, and (e) _____.

8a. _____

8b. _____

8c. _____

8d. _____

8e. _____

9. The timer output instruction is energized when the (a) _____ time equals the (b) _____ time.

9a. _____

9b. _____

10. Loss of power or logic continuity to a nonretentive on-delay timer will automatically reset the accumulated time to zero. (True or False)

10. _____

11. If the preset time of a timer is 100 and the time base is 0.1 s, the time-delay period would be _____ s.

11. _____

12. A (n) _____ timer must be intentionally reset with a separate signal.

12. _____

13. The programming of two or more timers together is called _____.

13. _____

14. The retentive timer reset (RES) instruction is always given the same address as the timer it resets. (True or False)

14. _____

15. Match each of the programmed timer instructions to the most direct circuit application. Place the number from the Timer Instructions list in the answer column.

TIMER INSTRUCTIONS
1) on-delay nonretentive timer
2) on-delay retentive timer
3) off-delay nonretentive timer

CIRCUIT APPLICATION
a) An alarm is to be switched on whenever a piping system has sustained a cumulative overpressure of 60 s.
b) A lamp is to be switched on 10 s after a switch has been actuated from the off to the on position.
c) A lamp is to be switched off 10 s after a switch has been actuated from the on to the off position.

15a. _____

15b. _____

15c. _____

16. When a time-delay period longer than the maximum preset time allowed for a single timer is required, the problem can be solved by programming two or more timers together. (True or False)

16. _____

17. Normally, the reset input to a timer will override the control input of the timer. (True or False)

17. _____

18. A retentive timer must be completely timed out to be reset. (True or False)

18. _____

19. A retentive timer will reset to 0 when the timer is de-energized. (True or False)

19. _____

20. The instantaneous contacts of a timer have no time-delay period associated with them. (True or False)

20. _____

21. Allen-Bradley timers can be located anywhere on the data organization table. (True or False)

21. _____

22. What timer instruction (TON, TOF, or RTO) would be best suited for each of the following control application:
a) Keep track of the total time to make one batch of product, even if the process is halted and then started again.

22a. _____

b) Hold the clamp on for 25 s after the glue is applied.

22b. _____

c) Open a valve 27 s after a switch is turned on. If interrupted, the valve should close and the time should reset to 0.

22c. _____

d) Begin timing when the rung is true and hold the accumulated time when rung logic goes false.

22d. _____

23. The accumulated time of a TOF timer is reset by causing the rung to go true momentarily. (True or False)

23. _____

PROGRAMMING ASSIGNMENTS FOR CHAPTER 7

1. Prepare an I/O connection diagram and ladder logic program for a nonretentive on-delay timer that will turn a light on 5 s after a switch is closed. Enter the program into the PLC and prove its operation.

2. Prepare an I/O connection diagram and ladder logic program for a nonretentive off-delay timer that will turn a light off 10 s after a switch is opened. Enter the program into the PLC and prove its operation.

3. Prepare an I/O connection diagram and ladder logic program to execute the hardwired timer circuit in Fig. 7-11. Enter the program into the PLC and prove its operation.

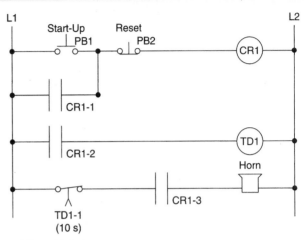

Fig. 7-12

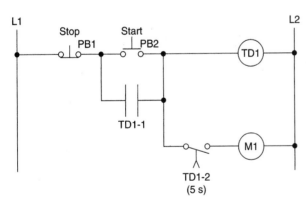

Fig. 7-11

4. Prepare an I/O connection diagram and ladder logic program to execute the hardwired start-up warning signal circuit in Fig. 7-12. Enter the program into the PLC and prove its operation.

5. Prepare an I/O connection diagram and ladder logic program to execute the hardwired automatic sequential control system in Fig. 7-13. Enter the program into the PLC and prove its operation.

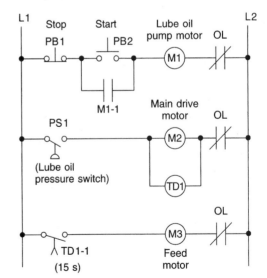

Fig. 7-13

6. Prepare an I/O connection diagram and ladder logic program to execute the hardwired pneumatic off-delay timer circuit in Fig. 7-14. Enter the program into the PLC and prove its operation.

7. Prepare an I/O connection diagram and ladder logic program for a retentive on-delay timer that will turn a light on anytime a switch is closed for an accumulated time of 15 s. Enter the program into the PLC and prove its operation.

8. Prepare an I/O connection diagram and ladder logic program to execute the hardwired sequential time delayed motor starting circuit in Fig. 7-15. Enter the program into the PLC and prove its operation.

9. Construct the annunciator flasher program in Fig. 7-16 using any available addresses, switches, and lights on your PLC demonstration panel. After constructing your program on paper, enter it into the PLC and prove its operation.

10. Prepare an I/O connection diagram and ladder logic program for two timers cascaded to give you a longer time-delay period than the maximum preset time allowed for the single timer. Enter the program into the PLC and prove its operation.

11-1. Outline the operating sequence for the hardwired timer-control circuit in Fig. 7-17.

11-2. Prepare an I/O connection diagram and ladder logic program that will correctly execute the circuit in Fig. 7-17. Enter the program into the PLC and prove its operation.

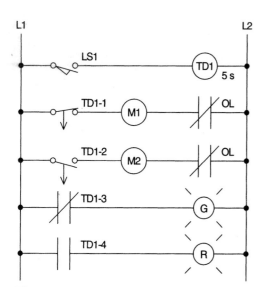

Fig. 7-14

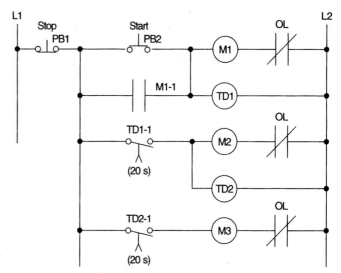

Fig. 7-15

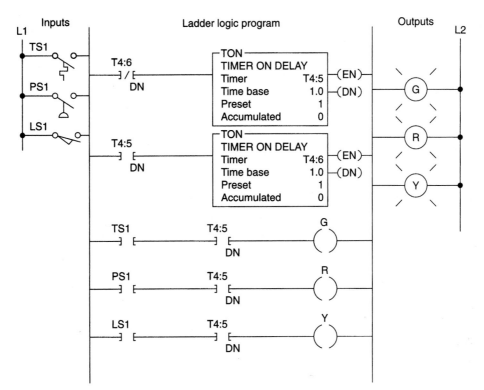

Fig. 7-16

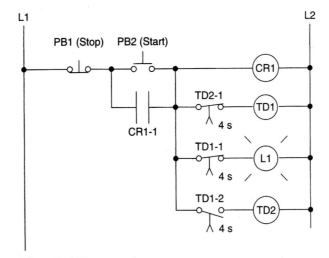

Fig. 7-17

12. Prepare an I/O connection diagram and ladder logic program for the simplified automatic mixing process in Fig. 7-18a from the given description of the process and existing relay schematic diagram in 7-18b. Enter the program into the PLC and prove its operation.

Process flow diagram

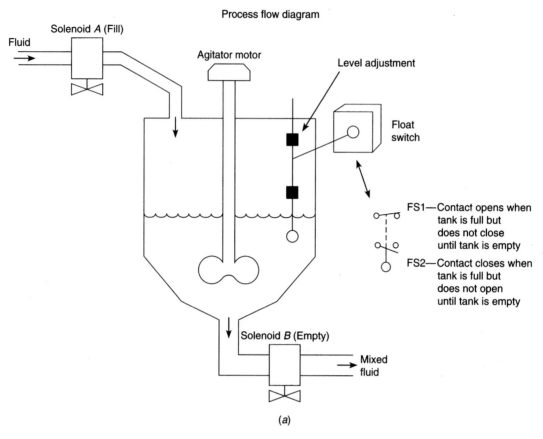

(a)

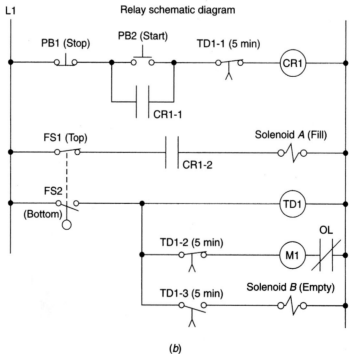

(b)

Fig. 7-18

OPERATIONAL SEQUENCE
1. Process is initiated by pressing the start pushbutton PB2.
2. Solenoid *A* is energized to allow fluid to flow into the tank.
3. Fluid flows in until the float switch is activated at the full position.
4. Agitator motor is started and operates for 5 min and then stops.

5. Solenoid *B* is energized to empty the tank.
6. When the float switch is activated at the empty position, the process stops and is placed in the ready position for the next manual start.

13. Prepare an I/O connection diagram and ladder logic program for the hardwired relay schematic diagram in Fig. 7-19. Enter the program into the PLC and prove its operation.

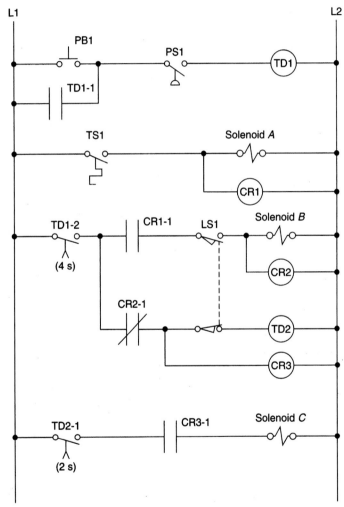

Fig. 7-19

14. A control process consists of three motors: M1, M2, and M3. The electrical control system is to be designed so that motor M1 must be running before motor M2 or M3 can be started. Each motor has its own start/stop pushbutton station. Both motors M2 and M3 can normally be stopped or started without affecting the operation of motor M1. However, if *all three* motors are running, the stopping of any one motor, for any reason, will automatically stop all three motors. Prepare an I/O connection diagram and ladder logic program from the relay schematic diagram (shown in Fig. 7-20) used to implement this control sequence. Enter the program into the PLC and prove its operation.

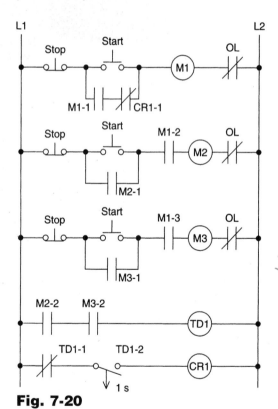

Fig. 7-20

15. Write a PLC program that will simulate the operation of the dc resistance motor starter circuit shown in Fig. 7-21. Assume timer presets of 5 s each. Enter the program into the PLC and prove its operation.

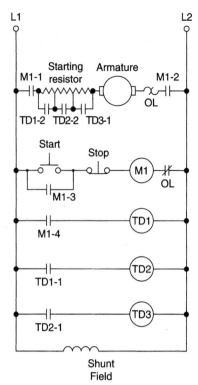

Fig. 7-21

16. Design a program to implement the fluid pumping process described in Chapter 7 of the text (Fig. 7-20). Enter the program into the PLC and prove its operation.

17. Design a program to implement the control of traffic lights in one direction as described in Chapter 7 of the text (Fig. 7-28). Enter the program into the PLC and prove its operation.

18. a) Design a program to implement the control of traffic lights in two directions as described in Chapter 7 of the text (Fig. 7-29). Enter the program into the PLC and prove its operation.
   b) Modify the program so there is a 2-s period when both directions will have their red lights illuminated. Enter the program into the PLC and prove its operation.

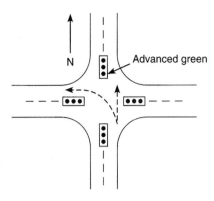

(a) Sketch of intersection

	10 s	20 s	5 s	20 s	5 s
Green north light	▓	██			
Green south light		██			
Yellow north/south light			██		
Red north light				██	██
Red south light	██			██	██
Green east/west light				██	
Yellow east/west light					██
Red east/west light	██	██	██		

(b) Timing chart

Fig. 7-22

19. Design a traffic light program consisting of two sets of lights controlling traffic as illustrated in Fig. 7-22.

- One set of lights controls traffic in a north/south direction; the other controls traffic in an east/west direction.
- The north lights have an advanced green that flashes at 1-s intervals for 10 s. It then changes to solid green for 20 s, yellow for 5 s, and red for 25 s. The south lights have no advanced green, so the south light is red for 35 s.

20. Write a PLC ladder logic diagram for a display sign that will sequentially turn on three lights 2 s apart, then turn all three lights off and repeat the sequence. Enter the program into the PLC and prove its operation.

21. Write a PLC program that will turn on pilot light PL1 10 s after switch S1 is turned on. Pilot light PL2 will come on 5 s after PL1 comes on. Pilot light PL3 will come on 8 s after PL2 comes on. Pressing pushbutton PB1 will reset all the timers but only if PL3 is on. Enter the program into the PLC and prove its operation.

22. When a switch is turned on, PL1 goes on immediately and P12 goes on 9 s later. Opening the switch turns both lights off. Write a program that implements this process. Enter the program into the PLC and prove its operation.

23. When a switch is turned on, PL1 and PL2 immediately go on. When the switch is turned off, PL1 immediately goes off. PL2 remains on for another 3 s and then goes off. Write a program that will implement this process. Enter the program into the PLC and prove its operation.

24. When a switch is turned on, PL1 and PL2 immediately go on. PL1 turns off after 4 s. PL2 remains on until the switch is turned off. Turning the switch off at any time turns both lights off. Write a program that will implement this process. Enter the program into the PLC and prove its operation.

25. A saw, fan, and lube pump all go on when a start button is pressed. Pressing a stop button immediately stops the saw but allows the fan to continue operating. The fan is to run for an additional 5 s after shutdown of the saw. If the saw has operated for more than 20 s, the fan should remain on until reset by a separate fan reset button. If the saw has operated less than 20 s, the lube pump should go off when the saw is turned off. However, if the saw has operated for more than 20 s, the lube pump should remain on for an additional 10 s after the saw is turned off. Write a program that will implement this process. Enter the program into the PLC and prove its operation.

26. Modify the continuous filling operation program described in Chapter 6 (Example 6-3 and Fig. 6-60) of the text. The modification calls for:
- A 2-s time delay period preceding the filling of the box after the conveyor has stopped.
- A 4-s time delay period preceding the starting of the conveyor after the box is filled.

Enter the program into the PLC and prove its operation.

8 PROGRAMMING COUNTERS

TEST 8•1

Choose the letter that best completes the statement.

Answer

1. Programmed counters can:
   a) count up.
   b) count down.
   c) be combined to count up and down.
   d) all of the above.

1. D

2. The counter instruction is found on:
   a) practically every PLC model.
   b) small-size PLCs.
   c) medium-size PLCs.
   d) large-size PLCs.

2. A

3. The PLC counter instruction is similar to the:
   a) internal relay instruction.
   b) transitional contact instruction.
   c) relay coil and contact instruction.
   d) timer instruction.

3. D

4. The output of a PLC counter is energized when the:
   a) accumulated count equals the preset count.
   b) preset count is greater than the accumulated count.
   c) count input rung is true.
   d) count input rung is false.

4. A

5. Which of the following is *not* usually associated with a PLC counter instruction?
   a) Address
   b) Preset value
   c) Time base
   d) Accumulated value

5. C

6. A CTU counter counts:
   a) scan transitions.
   b) true-to-false transitions.
   c) false-to-true transitions.
   d) both *b* and *c*.

6. C

7. The RES instruction:

   a) is used to reset the counter.

   b) is given the same reference address as the counter instruction.

   c) decrements the count when actuated.

   d) both *a* and *b.*

7. _D_

8. For the PLC counter to reset, the counter reset rung must:

   a) be true.

   b) be false.

   c) be either true or false, depending on the manufacturer.

   d) undergo a true-to-false transition.

8. _C_

9. Normally counters are retentive. This means that if your accumulated count is up to 0300 and power to your system is lost, when power is restored the accumulated count will be:

   a) 0000.

   b) 0250.

   c) 0300.

   d) 9999.

9. _C_

10. A one-shot, or transitional, contact:

   a) operates the same as an NO contact instruction.

   b) operates the same as an NC contact instruction.

   c) operates the same as a timed closed contact.

   d) closes for only one program scan when actuated.

10. _D_

11. A CTD counter counts:

   a) scan transitions.

   b) true-to-false transitions.

   c) false-to-true transitions.

   d) both *b* and *c.*

11. _C_

12. The accumulated count of a CTD counter:

   a) increments with each true-to-false transition.

   b) decrements with each true-to-false transition.

   c) decrements with each false-to-true transition.

   d) increments with each false-to-true transition.

12. _C_

13. A counter is to be programmed to keep track of the number of parts coming off a production line. If you wanted to subtract the number of rejected parts so your counter would count only the good parts, you would program:

   a) two up-counters.

   b) two down-counters.

   c) an up/down-counter.

   d) a counter with a transitional contact input.

13. _C_

14. In the program of Fig. 8-1, output PL2 will be energized: **14.** _____B____
    a) until the accumulated value equals the preset value.
    b) when the accumulated value equals the preset value.
    c) only when the accumulated value exceeds 010.
    d) only when the accumulated value is zero.

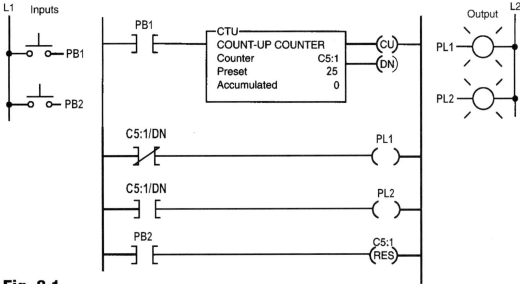

Fig. 8-1

15. In the program of Fig. 8-1, output PL1 will be energized: **15.** _____A____
    a) until the accumulated value equals the preset value.
    b) when the accumulated value equals the preset value.
    c) only when the accumulated value is less than 010.
    d) only when the accumulated value is 999.

16. In the program of Fig. 8-1, the instruction that will   **16.** _____
cause the counter to increment is:
    a) input PB1.
    b) input PB2.
    c) output PL1.
    d) output PL2.

17. In the program of Fig. 8-2, output Lot Full Light will be energized   **17.** _____
when the accumulated count is:
    a) 000.
    b) 025.
    c) 010.
    d) 020.

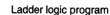

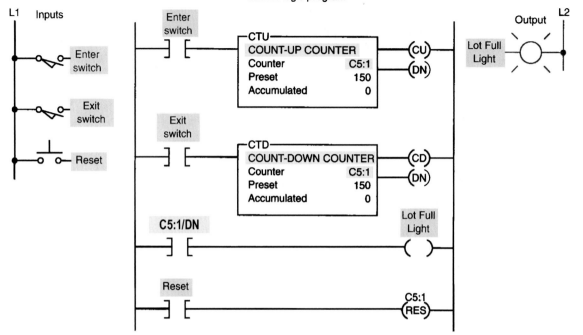

Fig. 8-2

18. In the program of Fig. 8-2, what is the instruction that will cause the counter to decrement?
a) Enter switch input.
b) Exit switch input.
c) C5:1 done bit.
d) Reset input.

18. _____

19. In the program of Fig. 8-2, what is the instruction that will preset the counter to a count of 000 when true?
a) Enter switch input.
b) Exit switch input.
c) C5:1 done bit.
d) Reset input.

19. _____

20. In the program of Fig. 8-2, suppose the accumulated count is 060 before the Enter switch input is actuated 15 times and the Exit switch input is actuated 5 times. After this operational sequence, the accumulated count would be:
a) 080.
b) 065.
c) 075.
d) 070.

20. _____

21. In the program of Fig. 8-2, during normal operation the accumulated value of CTU would always be:
a) the same as that of CTD.
b) 150.
c) 000 to 150.
d) 300.

21. _____

22. In the program of Fig. 8-2, suppose the accumulated count is 100. The following order of events then occurs: Exit switch input is actuated 20 times, Reset input is actuated 10 times, and Enter switch input is actuated 5 times. After this sequence, the accumulated count would be:
   a) 100.
   b) 115.
   c) 005.
   d) 000.

22. _____

23-1. The program of Fig. 8-3 is designed to:
   a) count up and count down.
   b) record the time of an event.
   c) count beyond the maximum count allowed per counter.
   d) count below the maximum count allowed per counter.

23-1. _____

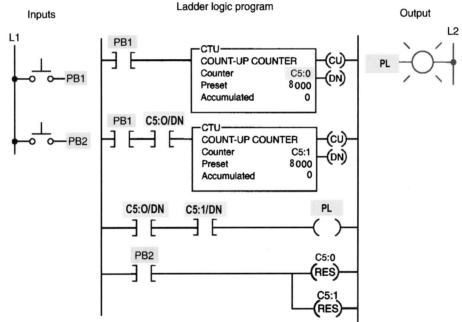

Fig. 8-3

23-2. Counter C5:1 starts counting:
   a) when the accumulated value of C5:0 reaches 8000.
   b) whenever input PB1 is actuated.
   c) whenever input PB2 is actuated.
   d) either *b* or *c*.

23-2. _____

23-3. Output PL will be energized when:
   a) the accumulated value of C5:0 reaches 8000.
   b) counter C5:0 is reset.
   c) the accumulated value of C5:1 reaches 8000.
   d) the accumulated value of C5:0 and C5:1 reaches 8000.

23-3. _____

23-4. When output PL is energized, how many counts have occurred?
    a) 4000
    b) 8000
    c) 16,000
    d) 99,999

23-4. _____

23-5. If you wanted output PL to go on after a count of 14,000, you would change the preset count of C5:1 to:
    a) 9999.
    b) 6000.
    c) 4000.
    d) 1000.

23-5. _____

23-6. When input PB2 is actuated:
    a) output PL is switched off.
    b) counter C5:0 is reset.
    c) counter C5:1 is reset.
    d) all of the above.

23-6. _____

24-1. The 24-h-clock program of Fig. 8-4 uses:
    a) 3 timers and 3 counters.
    b) 2 timers and 2 counters.
    c) 1 timer and 2 counters.
    d) 2 timers and 1 counter.

24-1. _____

24-2. Counter C5:1 is preset for:
    a) 12.
    b) 24.
    c) 60.
    d) 120.

24-2. _____

24-3. Counter C5:0 is preset for:
    a) 12.
    b) 24.
    c) 60.
    d) 120.

24-3. _____

24-4. RTO is preset for a:
    a) 60-s time period.
    b) 2-min time period.
    c) 12-h time period.
    d) 24-h time period.

24-4. _____

24-5. A false-to-true of rung No. 002 increments the clock by:
    a) 1 ms.
    b) 1 s.
    c) 1 min.
    d) 1 h.

24-5. _____

Ladder logic program

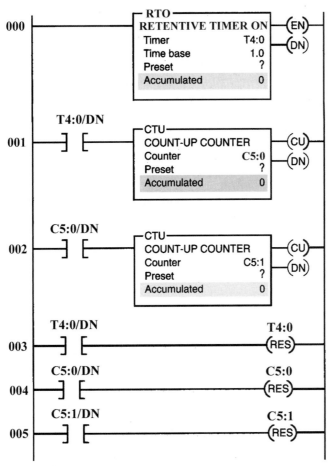

Fig. 8-4

24-6. Rung No. 003 undergoes a true-to-false transition once every:
    a) 60 s.
    b) 2 min.
    c) hour.
    d) 24 h.

24-6. _____

24-7. Assume the accumulated count of Counter C5:1 is 14 and that of C5:0 is 10. The correct time of day would be:
    a) 2:10 p.m.
    b) 10:14 a.m.
    c) 10:14 p.m.
    d) 2:10 a.m.

24-7. _____

25. In an up-counter, the _____ is incremented when the rung associated with the counter goes from false to true.
    a) preset value
    b) accumulated value
    c) reset
    d) counter enable

25. _____

26. When the up-counter reset is set to true, the following happens: **26.** _____
    a) the preset value is set to 0.
    b) the reset is set to false.
    c) the accumulated value is set to 0.
    d) the accumulated value is set to maximum.

27. In an up-counter, when the accumulated count exceeds the preset   **27.** _____
count without a reset, the accumulated count will:
    a) set itself to 0.
    b) start decrementing.
    c) continue incrementing.
    d) hold the accumulated value.

28. When the accumulated count exceeds the preset count, the: **28.** _____
    a) accumulated value is set to 0.
    b) preset is set to 0.
    c) reset changes state.
    d) counter done bit is true.

TEST 8 • 2

Place the answers to the following questions in the answer column at the right.

Answer

1. Programmed counters can serve the same function as mechanical counters. (True or False)

1. _____

2. The majority of counters used in industry are classified as _____ counters.

2. _____

3. Practically every PLC model offers some form of counter instruction as part of its instruction set. (True or False)

3. _____

4. Counters are similar to timers, except that they do not operate on an internal clock. (True or False)

4. _____

5. Counter instructions can be (a) _____ formatted or (b) _____ formatted.

5a. _____

5b. _____

6. The up-counter increments its accumulated value by 1 each time the counter rung makes a(n) _____ transition.

6. _____

7. The output of the counter is energized whenever the accumulated count is less than or equal to the preset count. (True or False)

7. _____

8. A programmed counter is reset by means of a counter _____ instruction.

8. _____

9. Most PLC counters are normally nonretentive. (True or False)

9. _____

10. Some PLC counters operate on the leading edge of the input signal, while others operate on the trailing edge. (True or False)

10. _____

11. All PLC manufacturers require the reset rung or line to be true to reset the counter. (True or False)

11. _____

12. A transitional off-to-on contact will allow logic continuity for one scan and then open, even though the triggering signal may stay on. (True or False)

12. _____

13. The transitional contact instruction is also known as a(n) _____ contact instruction.

13. _____

14. Transitional contacts are often used for _____ counters and timers.

14. _____

15. A down-counter output instruction will decrement by 1 each time the counted event occurs. (True or False)

15. _____

16. In normal use, the down-counter is used in conjunction with the up-counter to form an up/down-counter. (True or False)

16. _____

17. All up-counters count only to their preset values, and additional counts are ignored. (True or False)

17. _____

18. One way of counting events that exceed the maximum number allowable per counter instruction is by the _____ of two counters.

18. _____

19. The logic used to implement a simple clock as part of a PLC's program involves the use of a single timer instruction along with a pair of counter instructions. (True or False)

19. _____

20. When programming a counter coil reset (RES), it is given the same reference address as the counter it is to reset. (True or False)

20. _____

21. The counter enable bit is true whenever the counter instruction is false. (True or False)

21. _____

22. The counter done bit is true whenever the (a) _____ value is equal to or greater than the (b) _____ value.

22a. _____

22b. _____

23. The counter _____ bit is true whenever the counter counts past its maximum value.

23. _____

24. The counter _____ values specifies the value that the counter must count to before it changes the state of the done bit.

24. _____

25. The counter accumulated value is the current count based on the number of times the rung goes from false to true. (True or False)

25. _____

26. The counter number C5:4 represents counter file 5, counter 4 in that file. (True or False)

26. _____

27. Encoder pulses can be counted to measure distance. (True or False)

27. _____

PROGRAMMING ASSIGNMENTS
FOR CHAPTER 8

1. Construct the up-counter program in Fig. 8-5 using any available addresses, switches, and lights on your PLC demonstration panel. After constructing your program on paper, enter it into the PLC and prove its operation.

Ladder logic program

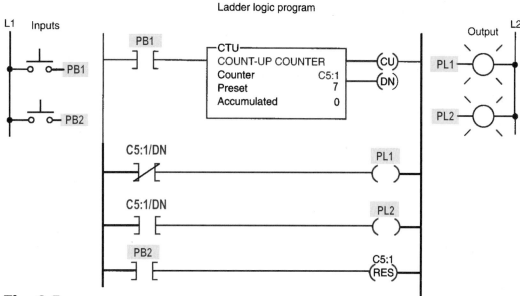

Fig. 8-5

2. Design a PLC program and prepare an I/O connection diagram and ladder logic program that will correctly execute the conveyor motor control process in Fig. 8-6. Enter the program into the PLC and prove its operation.

OPERATIONAL SEQUENCE
1. The start button is pressed to start the conveyor motor.
2. Cases move past the proximity switch and increment the counter's accumulated value.

3. After a count of 50, the conveyor motor stops automatically and the counter's accumulated value is reset to 0.
4. The conveyor motor can be stopped and started manually at any time with out loss of the accumulated count.
5. The accumulated count of the counter can be reset manually at any time by means of the count reset button.

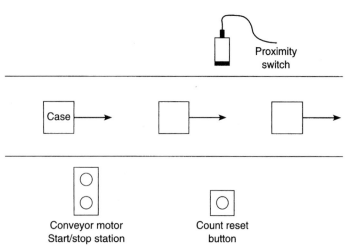

Process flow diagram

Proximity switch

Case

Conveyor motor
Start/stop station

Count reset
button

Fig. 8-6

3. Design a PLC program and prepare an I/O connection diagram and ladder logic program to execute the following alarm monitor circuit. Enter the program into the PLC and prove its operation.

OPERATIONAL SEQUENCE FOR ALARM MONITOR
1. An alarm light is triggered by the closing of a liquid level switch (LS1).
2. The light will *flash* whenever the alarm condition is triggered and has not been acknowledged, even if the alarm condition clears in the meantime.
3. The alarm is acknowledged by closing a selector switch (SS1).
4. The light will operate in the *steady ON* mode when the alarm trigger condition still exists but has been acknowledged.

4. Construct the up/down-counter program in Fig. 8-7 using any available addresses, switches, and lights on your PLC demonstration panel. After constructing your program on paper, enter it into the PLC and prove its operation.

5. Design a PLC program and prepare an I/O connection diagram and ladder logic program to execute the following parking garage counter requirements. Enter the program into the PLC and prove its operation.

OPERATIONAL SEQUENCE FOR THE PARKING GARAGE COUNTER
1. An up-counter is used to keep a running tally on the total number of cars that enter.
2. An up/down-counter is used to keep track of the total number of cars in the parking garage at any one time.
3. Any time a total of 50 cars is in the garage, the Lot Full light automatically comes on.
4. Two reset buttons are provided for manually resetting the accumulated count of each counter.

6. Design a PLC program that will allow you to count events that exceed the maximum number allowable per counter instruction by the cascading of two counters. After constructing your program on paper, enter it into the PLC and prove its operation to the instructor.

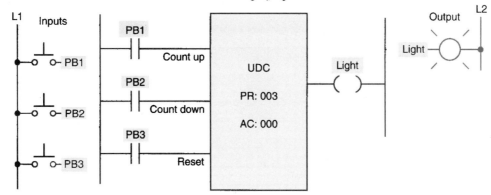

Fig. 8-7

7. Construct a simple timer/counter program that produces a time-of-day clock measuring time in hours and minutes. After constructing your program on paper, enter it into the PLC and prove its operation.

8. Design a PLC program to execute the following monitoring of the time of an event using the 24-h-clock program of Question 7. After constructing your program on paper, enter it into the PLC and prove its operation.

OPERATIONAL SEQUENCE
1. The time of the opening of a pressure switch is to be monitored.
2. The circuit is set into operation by pressing a reset button and setting the programmed clock for the current time of day.
3. A set pilot light comes on to indicate that the clock is operating and the circuit is set to operate.
4. Should the pressure switch open at any time, the clock will stop automatically, the set pilot light will switch off, and a second trip pilot light will switch on.
5. The clock is then read to determine the time of opening of the pressure switch.

9. Design a PLC program and prepare an I/O connection diagram and ladder logic program that will correctly execute the packaging process in Fig. 8-8. Enter the program into the PLC and prove its operation.

OPERATIONAL SEQUENCE OF PACKAGING PROCESS
1. The purpose of this process is to deposit 50 pieces of the product in each container.
2. The process is set in operation by pressing a start pushbutton.
3. As the product passes through the light beam, it is detected by the photoelectric proximity switch and counted by the PLC counter.
4. When the count reaches 50, the solenoid-operated deflector plate (SOL *A*) energizes to channel the product from chute A to chute B.
5. The counter is reset automatically for the next count of 50.
6. When the second count of 50 is reached, the solenoid-operated deflector plate de-energizes to channel the product back into chute A, and so on.
7. Provisions are also made for stopping the process at any time and manually resetting the accumulated value of the counter to any number.

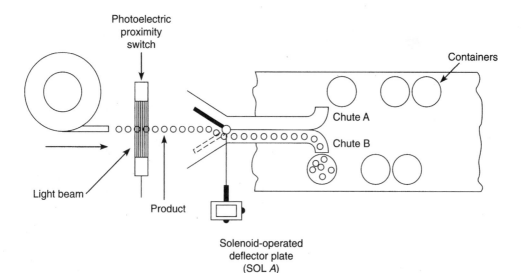

Photoelectric proximity switch

Containers

Chute A

Chute B

Light beam

Product

Solenoid-operated deflector plate (SOL *A*)

Fig. 8-8

10. Design a program that will latch on an output, PL1, after an input, PB1, has cycled on 20 times. When the count of 20 is reached, the counter will reset itself automatically. PB2 will unlatch PL1. Enter the program into the PLC and prove its operation.

11. Design a program to implement the parts-counting process described in Chapter 8 of the text (Fig. 8-13). Enter the program into the PLC and prove its operation.

12. Design a program to implement the up/down-counter described in Chapter 8 of the text (Fig. 8-22). Enter the program into the PLC and prove its operation.

13. Design a program to implement the in-process monitoring system described in Chapter 8 of the text (Fig. 8-23). Enter the program into the PLC and prove its operation.

14. Design a program to implement the cascading of counters described in Chapter 8 of the text (Fig. 8-25). Enter the program into the PLC and prove its operation.

15. Design a program to implement the length measurement process described in Chapter 8 of the text (Fig. 8-30). Enter the program into the PLC and prove its operation.

16. Design a program to implement the automatic stacking process described in Chapter 8 of the text (Fig. 8-31). Enter the program into the PLC and prove its operation.

17. Design a program to implement the motor lock-out process described in Chapter 8 of the text (Fig. 8-32). Enter the program into the PLC and prove its operation.

18. Design a program to implement the product flow-rate process described in Chapter 8 of the text (Fig. 8-33). Enter the program into the PLC and prove its operation.

19. Design a program to implement the timer driving a counter described in Chapter 8 of the text (Fig. 8-34). Enter the program into the PLC and prove its operation.

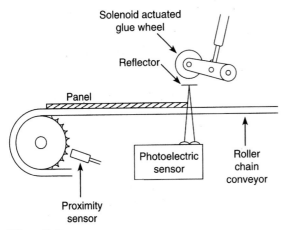

Solenoid actuated glue wheel

Reflector

Panel

Photoelectric sensor

Roller chain conveyor

Proximity sensor

Fig. 8-9

20. Design a program to implement the glue application process illustrated in Fig. 8-9. The sequence of operation is as follows:
    1. Two up-counters are used to control the glue application on structural panels.
    2. Glue is applied in a band that starts a fixed distance from the leading end of the panel and runs for a predetermined length along the panel.
    3. Panels are moved past the gluing station by a 1-in. pitch roller chain conveyor.
    4. Count pulses are generated by sensing sprocket teeth with a proximity sensor.
    5. A photoelectric sensor keeps the counter reset to 0 as long as it senses reflected light through a clear track in the chain conveyor. When the leading edge of a panel enters the glue application point, the light beam is broken, the reset condition is removed, and the counter begins accumulating counts.
    6. When the count reaches the preset of the first counter, the glue wheel is actuated, starting the glue strip.
    7. When the count reaches the preset of the second counter, the glue wheel is de-actuated.

8. When the trailing end of the panel leaves the field of view of the photoscanner, the counter is reset to 0 until the arrival of the next panel.

21. Write a PLC program that will simulate the operation of the hard-wired relay control circuit shown in Fig. 8-10 on p. 132. The sequence of operation is as follows:
    1. The system is initiated by an operator each shift, and the items to be sorted are then fed onto the production line conveyor.
    2. Once on the conveyor, the items proceed, operating limit switch 1, which counts all the items.
    3. Limit switch 2 counts only the larger items.
    4. As limit switch 2 is operated, a pneumatic ram is activated and thus stores all the larger items in packing box 2.
    5. The smaller items continue to the end of the conveyor and are stored in packing box 1.
    6. At the end of the shift, when the operator presses the end-of-work button, the conveyor continues to run for 2 min until all the items currently on the conveyor are cleared and packed.

22. Design a program to implement the stacking/painting process illustrated in Fig. 8-11 on p. 132. The sequence of operation is as follows:
    1. When the start button is pressed, a stacker starts stacking sheets at position A.
    2. After 6 sheets are stacked, the stacker stops, and the conveyor goes on and moves the stack to position B.
    3. At position B, paint is applied through a solenoid valve for 10 s.
    4. After painting is complete, the conveyor is restarted manually and moves the stack to position C.

5. At position C, the stack stops automatically and the stack is removed manually.
6. A stop button stops the process any time it is depressed.
7. Assume that only one stack is on the conveyor at a time, and include whatever sensors may be required.

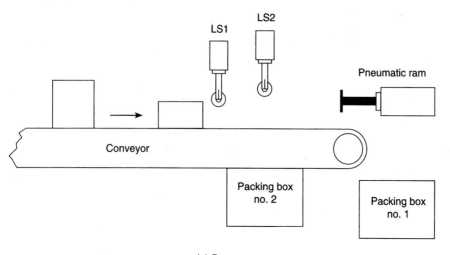

(a) Process

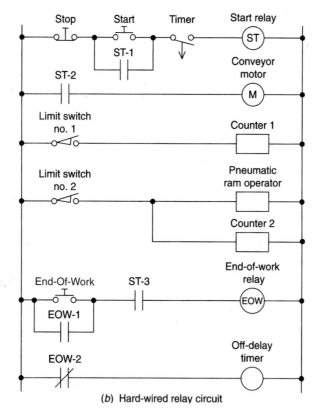

(b) Hard-wired relay circuit

Fig. 8-10

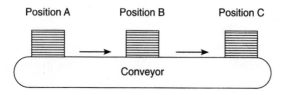

Fig. 8-11

23. Write a program designed to alternate the use of two input pumps so that they both get the same amount of usage over their lifetime. The operation of the process can be summarized as follows:

1. A start/stop pushbutton station is provided for control of the two input pump motors P1 and P2.
2. The start/stop pushbutton station is operated to control pump P1.
3. When the tank is full drain pump motor P3 is started automatically and runs until the low level sensor is actuated.
4. After 5 fillings of tank by pump P1 control automatically shifts to pump P2.
5. Operation of the start/stop pushbutton now controls pump P2.
6. After 5 fillings of the tank by pump P2 the sequence is repeated.

9 PROGRAM CONTROL INSTRUCTIONS

TEST 9•1

Choose the letter that best completes the statement.

Answer

1. Which of the following PLC instructions would *not* be classified as an override instruction?
   a) master control reset
   b) jump-to-subroutine
   c) output energize
   d) jump-to-label

1. C

2. The MCR instruction establishes a zone in the user program in which all nonretentive outputs can be:
   a) turned on simultaneously.
   b) turned off simultaneously.
   c) turned on in a defined sequence.
   d) turned off in a defined sequence.

2. B

3. In the program in Fig. 9-1, when the MCR instruction is false, output(s) _____ will *always* be de-energized.
   a) M, PL1 and SOL
   b) M and PL1
   c) PL1
   d) SOL

3. B

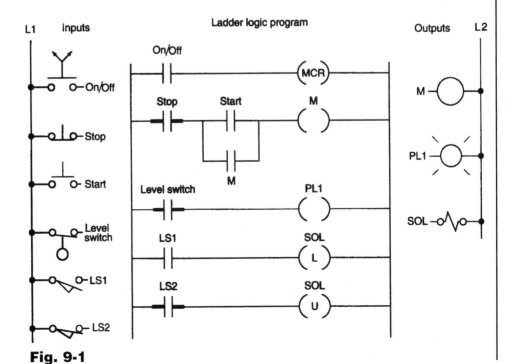

Fig. 9-1

4. In the program in Fig. 9-1, assume that the MCR instruction makes a false-to-true transition. As a result:
   a) all outputs will be controlled by the respective input conditions.
   b) all nonretentive outputs will de-energize.
   c) all retentive outputs will de-energize.
   d) all nonretentive outputs will energize.

4. _A_

5. In the program in Fig. 9-2, assume output SOL *C* is energized at the time the MCR instruction makes a true-to-false transition. As a result, output SOL *C* will:
   a) de-energize.
   b) remain energized.
   c) remain energized but still be controlled by inputs LS3 and LS4.
   d) de-energize but still be controlled by inputs LS3 and LS4.

5. _B_

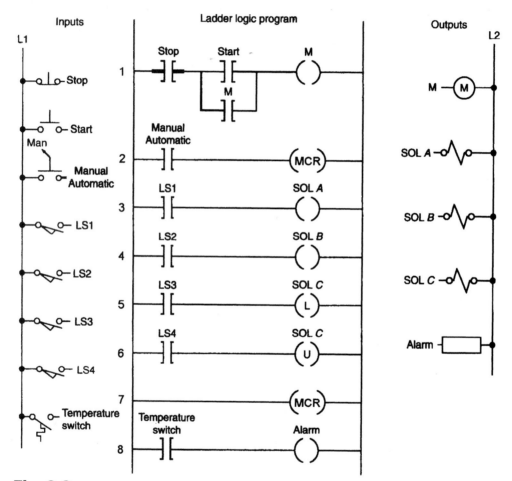

Fig. 9-2

6. In the program in Fig. 9-2, the fenced zone controlled by the MCR instruction is (are):
   a) rungs No. 1 through No. 8.
   b) rungs No. 3 through No. 6.
   c) rungs No. 2 and No. 6.
   d) rung No. 2.

6. _B_

7. In the program in Fig. 9-2, which of the following is the conditional instruction that controls the MCR zone?
a) Stop
b) Start
c) Manual Automatic
d) Temperature Switch

7. _C_

8. In the program in Fig. 9-2, the latch and unlatch instructions would be classified as:
a) retentive outputs.
b) nonretentive outputs.
c) condition instructions.
d) uncondition instructions.

8. _A_

9. In the program in Fig. 9-2, assume the alarm output is activated. This would require the:
a) temperature switch input to be true.
b) Manual Automatic input to be true.
c) LS4 input to be true.
d) both *a* and *b*.

9. _D_

10. The main advantage to the jump-to-label instruction is that:
a) any number of rungs may be programmed between the jump and label rungs.
b) it allows you to use one set of condition instructions to control multiple outputs.
c) it allows you to use one set of condition instructions to control multiple inputs.
d) it has the ability to reduce the processor scan time.

10. _D_

11. The label (LBL) instruction is:
a) always logically true.
b) has the same address as the jump instruction with which it is used.
c) is used to identify the ladder rung that is the target destination of the JMP instruction.
d) all of these.

11. _D_

12. Which of the following instructions would most likely be programmed outside the jumped area of a program?
a) Latch and unlatch instructions
b) Timer and counter instructions
c) Immediate inputs and outputs
d) Forced inputs and outputs

12. _____

13. In the program of Fig. 9-3, which output is *not* affected by the jump instruction?
a) M
b) SOL 1
c) SOL 2
d) PL2

13. _____

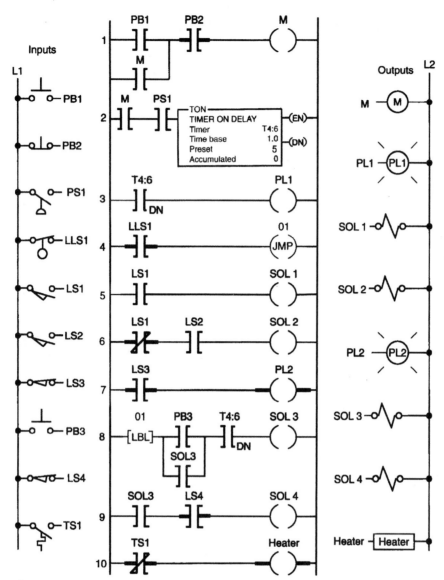

Fig. 9-3

14. In the program of Fig. 9-3, rungs 5, 6, and 7 are not scanned by the processor when rung _____ has logic continuity.
   a) 1
   b) 2
   c) 3
   d) 4

14. _____

15. In the program of Fig. 9-3, when the jump-to-label instruction is executed, the outputs of the jumped rungs:
   a) are all energized.
   b) are all de-energized.
   c) are all immediately updated.
   d) remain in their last state.

15. _____

16. The jump-to-subroutine instruction can save a great deal of duplicate programming in cases:

    a) that require the programming of several timers.

    b) that require the programming of several counters.

    c) where a machine has a portion of its cycle that must be repeated several times during one machine cycle.

    d) all of these.

16. _____

17. In the program of Fig. 9-4, when the examine on sensor instruction is true, the processor:

    a) turns on all outputs in the subroutine area.

    b) turns off all outputs in the subroutine area.

    c) stops executing the subroutine.

    d) starts executing the subroutine.

17. _____

Main Program
File 2

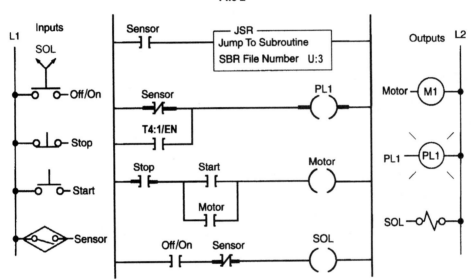

Subroutine
File 3

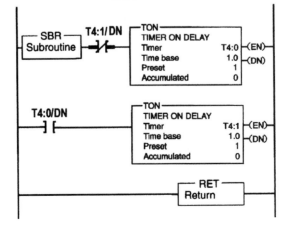

Fig. 9-4

18. In the program of Fig. 9-4, when the processor scan reaches the RET instruction, it will return the processor to the:
a) start of the program.
b) end of the program.
c) rung above the JSR instruction.
d) rung below the JSR instruction.

18. _____

19. The immediate input and output instructions provide a way of:
a) ending the program immediately.
b) restarting the program immediately.
c) temporarily interrupting the program scan to allow selected bits in the data table to be updated.
d) temporarily interrupting the program scan to reset all bits in the data table to 0.

19. _____

20. Immediate instructions should be used only when:
a) a program must be halted immediately.
b) a program must be restarted immediately.
c) the updating of an input or output is critical to your operation.
d) the resetting of all bits in the data table is critical to your operation.

20. _____

21. Immediate instructions are most useful when programmed:
a) immediately after the I/O scan has occurred.
b) immediately prior to the I/O scan.
c) at the middle or toward the end of the program.
d) near the beginning of the program.

21. _____

22. The use of immediate instructions:
a) increases the total scan time of the program.
b) decreases the total scan time of the program.
c) increases the number of rungs that can be programmed.
d) decreases the number of rungs that can be programmed.

22. _____

23-1. The forcing function of a PLC allows the user to turn an external input or output on or off:
a) according to the forced program.
b) according to the main program.
c) from the keyboard regardless of its actual status.
d) all of these.

23-1. _____

23-2. Forcing functions are often used:
a) to continue a machine process until a faulty field device can be repaired.
b) for testing purposes during an initial start-up.
c) for troubleshooting purposes.
d) all of these.

23-2. _____

23-3. In the program of Fig. 9-5, the actual status of input I:1/3 is _____ but the forced status is _____:
a) false . . . true
b) true . . . false
c) false . . . false
d) true . . . true

23-3. _____

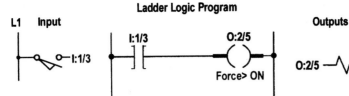

Fig. 9-5

23-4. In the program of Fig. 9-5, the output of O:2/5 would be _____
and the output of O:2/6 would be _____.
a) false . . . true
b) true . . . false
c) false . . . false
d) true . . . true

23-4. _____

23-5. In the program of Fig. 9-6, the actual status of output address O:2/5
is _____ , but the forced status is _____.
a) false . . . true
b) true . . . false
c) false . . . false
d) true . . . true

23-5. _____

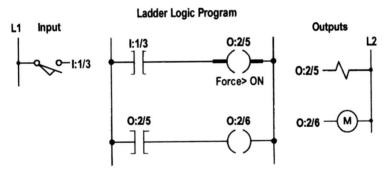

Fig. 9-6

23-6. In the program of Fig. 9-6, the status of examine on instruction O:2/5
would be _____ and the status of output O:2/6 would be _____.
a) false . . . true
b) true . . . false
c) false . . . false
d) true . . . true

23-6. _____

23-7. Forcing functions should *not* be used:
a) with retentive outputs.
b) with nonretentive outputs.
c) with immediate I/O instructions.
d) without consideration for any potential unsafe effects.

23-7. _____

24. PLC emergency stop circuits should be:
   a) hardwired outside the controller program.
   b) programmed as part of the master control reset instruction.
   c) programmed as part of the zone control last state instruction.
   d) programmed as an immediate input instruction.

24. _____

25-1. In the diagram in Fig. 9-7, power to the processor module is controlled by the:
   a) master control relay.
   b) start/stop buttons.
   c) main disconnect switch.
   d) all of the above.

25-1. _____

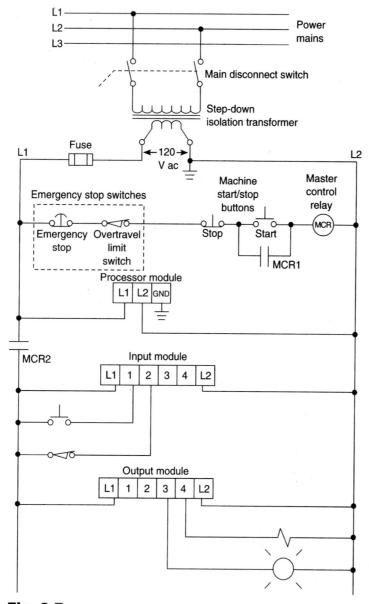

Fig. 9-7

25-2. In the diagram in Fig. 9-7, power to the input and output module is controlled by the:
a) master control relay.
b) start/stop buttons.
c) main disconnect switch.
d) all of the above.

25-2. _____

25-3. In the diagram in Fig. 9-7, the transformer is used to:
a) isolate the controller from the main power lines.
b) step up the main power line voltage.
c) provide the low-voltage dc operating voltage for the controller.
d) all of the above.

25-3. _____

25-4. In the diagram in Fig. 9-7, assume the processor comes equipped with a normally closed fault relay contact output designed to open when a processor malfunction is detected. This contact would be:
a) programmed as part of the master control reset instruction.
b) programmed as part of the zone control last state instruction.
c) hardwired in series with the emergency stop button.
d) hardwired in parallel with the emergency stop button.

25-4. _____

25-5. In the diagram in Fig. 9-7, when replacing modules or working on equipment controlled by the PLC installation, the safest way to proceed is to:
a) de-energize the MCR coil.
b) block open the emergency stop switch.
c) remove the fuse from the circuit.
d) pull and lock the disconnect switch.

25-5. _____

26-1. In the program in Fig. 9-8, the use of the starter auxiliary contact instead of a programmed contact:
a) is more costly.
b) is safer.
c) provides positive feedback about the exact status of the motor.
d) all of these.

26-1. _____

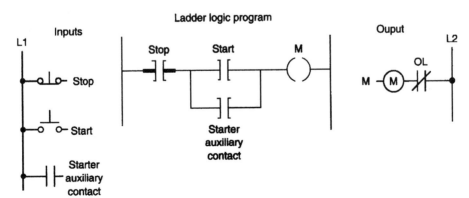

Fig. 9-8

26-2. In the program in Fig. 9-8, assume that the stop button was changed to a normally open contact type. As a result, the program could be made to operate as before by changing the instruction at address:
a) stop to examine if open.
b) start to examine if open
c) starter auxiliary contact to examine if open.
d) both *a* and *c*.

26-2. _____

26-3. In the program in Fig. 9-8, assume that the NC stop button is replaced with an NO stop button and that the program is changed so it operates as before. Should the field wire connected to one end of the stop button break off:
a) the motor would stop automatically.
b) pressing the stop button would stop the motor.
c) pressing both the start and stop buttons would stop the motor.
d) none of these.

26-3. _____

27. A jump instruction is similar to a(n):
a) MCR command.
b) ZCL command.
c) skip command.
d) JSR command.

27. _____

28. A JSR instruction:
a) tells the processor to jump from the main program to a subroutine area or file.
b) tells the process to execute the fault routine.
c) latches outputs when energized.
d) latches outputs when de-energized.

28. _____

29. The MCR instruction:
a) is an output instruction.
b) is used in pairs.
c) is used to disable or enable a zone within a ladder program.
d) all of these.

29. _____

30. The _____ is the target for the jump instruction.
a) LBL
b) TND
c) IOT
d) RET

30. _____

31. The _____ instruction will return the scan to your main program at the completion of the subroutine.
a) LBL
b) IIN
c) TND
d) RET

31. _____

32. The _____ instruction stops the processor from scanning the rest of the program.
a) LBL
b) IOT
c) TND
d) STI

32. _____

TEST 9•2

Place the answers to the following questions in the answer column at the right.

Answer

1. Master control reset (MCR) and jump (JMP) are often referred to as _____ instructions.

1. _____

2. The MCR instruction can only be programmed to control an entire circuit. (True or False)

2. _____

3. When the MCR instruction is _____, all rung outputs below the MCR will be controlled by their respective input conditions.

3. _____

4. If the MCR output is turned off or de-energized, all nonretentive rungs below the MCR will be

_____.

4. _____

5. _____ instructions should not normally be placed within an MCR zone because they will remain in their last active state when the instruction goes false.

5. _____

6. When programming an MCR instruction to control a fenced zone, an MCR rung with no conditional inputs is placed at the beginning of the zone and an MCR rung with conditional inputs is placed at the end of the zone. (True or False)

6. _____

7. The master control instruction is used as a substitute for a hardwired emergency stop switch. (True or False)

7. _____

8. The jump (JMP) instruction is used to jump over certain program instructions if certain conditions exist. (True or False)

8. _____

9. The advantage of the JMP instruction is the ability to reduce the processor _____.

9. _____

10. In a jump-to-label program, the _____ instruction is used to identify the ladder rung that is the target destination of the jump instruction.

10. _____

11. The label address number must match that of the jump instruction with which it is used. (True or False)

11. _____

12. The JMP instruction does not contribute to logic continuity and, for all practical purposes, is always logically true. (True or False)

12. _____

13. The jump-to-subroutine instruction is used where a machine has a portion of its cycle that must be _____ several times during one machine cycle.

13. _____

14. When the program scan reaches an immediate I/O instruction, the scan is interrupted and the bits of the addressed word are _____.

14. _____

15. The immediate I/O instruction is used with _____ I/O devices that require updating in advance of the I/O scan.

15. _____

16. The immediate I/O instruction is most useful if the instruction associated with the device is at the beginning of the program. (True or False)

16. _____

17. The use of the immediate I/O instruction increases the total _____ of the program.

17. _____

18. The forcing capability of a PLC allows the user to turn an external I/O on or off regardless of the _____ of the device.

18. _____

19. Random forcing of given inputs or outputs can cause equipment damage. (True or False)

19. _____

20. Emergency stop circuits should be _____ outside the controller so that, in the event of total controller failure, independent and rapid shutdown means are available.

20. _____

21. A main _____ is installed on the incoming power lines as a means of removing power from the entire PLC system.

21. _____

22. Power to the PLC input and output devices is usually controlled by means of a hard-wired _____ circuit.

22. _____

23. The master control relay can be used as a substitute for a disconnect switch. (True or False)

23. _____

24. The use of a motor starter seal-in contact in place of a programmed contact provides _____ feedback about the exact status of the motor.

24. _____

25. The safest way to wire a stop button to a PLC system is to use a(n) (a) _____ contact programmed to examine for a(n) (b) _____ condition.

25a. _____

25b. _____

26. The label instruction has a logical true condition. (True or False)

26. _____

27. Jumping to a subroutine does not cause any rungs of the main program to be skipped over. (True or False)

27. _____

28. The jump instruction allows a section of a program to be jumped when a production fault occurs. (True or False)

28. _____

29. It is not possible to jump backward in the program. (True or False)

29. _____

30. Nesting subroutines allow you to direct program flow from the (a) _____ program to a subroutine and then to another (b) _____.

30a. _____

30b. _____

31. Nested subroutines make complex programming easier. (True or False)

31. _____

32. Forcing outputs affects only the addressed output terminal. (True or False)

32. _____

33. Programming the selectable timed interrupt is done when a section of program needs to be executed on a(n) _____ basis rather than on an event basis.

33. _____

34. The fault routine allows for an orderly shutdown in case of a fault. (True or False)

34. _____

35. The temporary end instruction, when true, _____ the program scan.

35. _____

PROGRAMMING ASSIGNMENTS FOR CHAPTER 9

1. Construct a simulated program for the MCR program in Fig. 9-9 using any available addresses, switches, and lights on your PLC demonstration panel. After constructing your program on a separate sheet of paper, enter it into the PLC. Demonstrate that when the MCR instruction is de-energized, all nonretentive outputs de-energize and all retentive outputs remain in their last state.

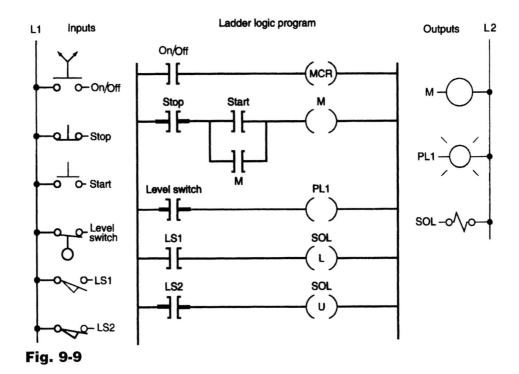

Fig. 9-9

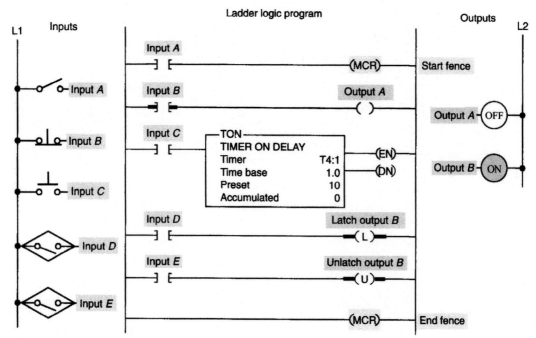

Fig. 9-10

2. Construct a simulated program for the MCR fenced zone program in Fig. 9-10. After constructing your program on a separate sheet of paper, enter it into the PLC. Demonstrate how the rungs between the two MCR instructions are controlled.

3. Construct a simulated program for the jump-to-label program in Fig. 9-11. After constructing your program on a separate sheet of paper, enter it into the PLC. Demonstrate how the jump-to-label instruction is executed.

4. Construct a simulated program for the jump-to-subroutine program in Fig. 9-12. After constructing your program on a separate sheet of paper, enter it into the PLC. Demonstrate how the jump-to-subroutine instruction is executed.

5. Enter the program in Fig. 9-13 into the PLC and demonstrate how each of the following is executed: (a) forcing the switch on; (b) forcing the switch off; (c) forcing PL1 on; and (d) forcing PL1 off.

Ladder logic program

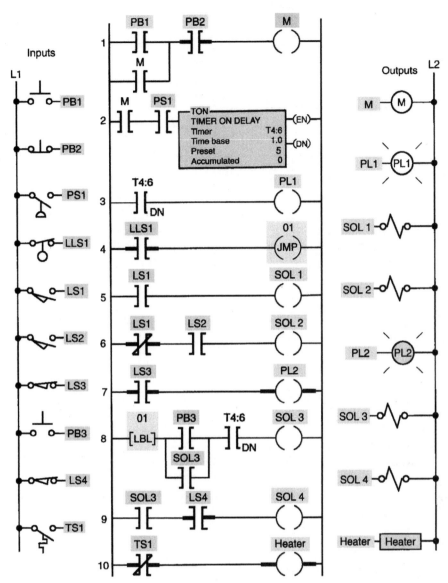

Fig. 9-11

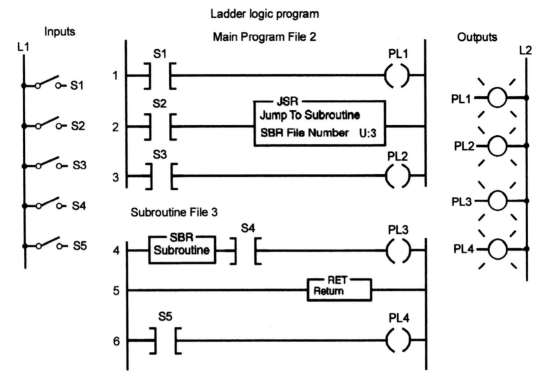

Fig. 9-12

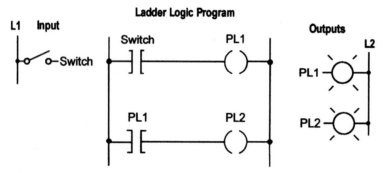

Fig. 9-13

6-1. Enter the simulated start/stop pushbutton program in Fig. 9-14 in the PLC and demonstrate to the instructor how an open in the stop pushbutton circuit will fail to de-energize the output.

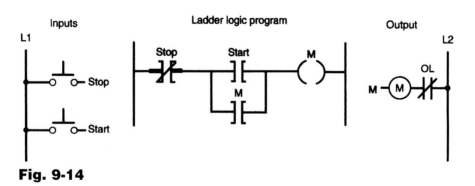

Fig. 9-14

6-2. Replace the normally open stop pushbutton with a normally closed type and modify the program so that the circuit operates properly. Demonstrate how an open in the stop pushbutton circuit of this program will automatically de-energize the output

7. Construct a simulated program for the MCR program of Fig. 9-15. Enter the program into the PLC. Operate the program according to the following sequence and answer the question(s) associated with each sequence:

a) Close switches 1, 2, 3, 4, and 6 and allow timer T4:2 to time out. What lights are on?

b) Open switch 1. What light is on now? Why did lights 1 and 2 go off?

c) Open switch 4 and close switch 5. Did light 1 go off? Why or why not?

d) What happened to the two timers when you disabled the MCR zone?

e) What happened to the two timers when you re-enabled the MCR zone?

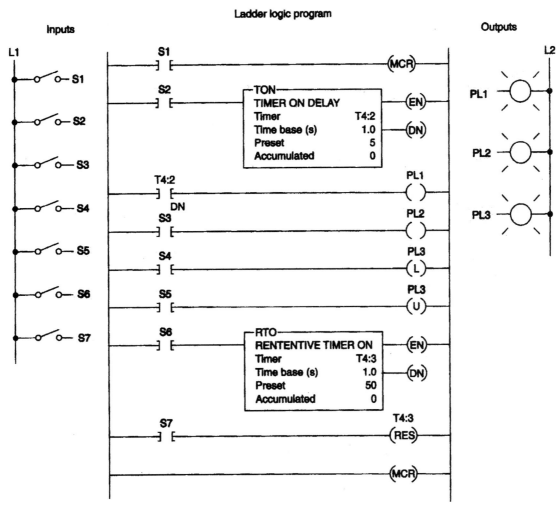

Fig. 9-15

Ladder logic program

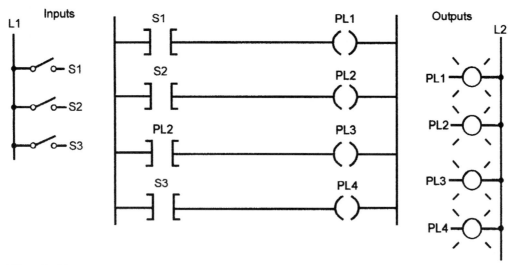

Fig. 9-16

8. Construct a simulated program for the ladder logic program of Fig. 9-16. Enter the program into the PLC. Operate the program to force specified inputs and outputs according to the following sequence:
   a) Turn switches 1 and 2 off and switch 3 on.
   b) Force off switch 3. Use the data monitor to observe the status of the corresponding bit for switch 3 in the input image table file. Close switch 3 and observe the status of the bit. How does forcing inputs manipulate the input image table file bits? Disable the force and exit from the data monitor to the ladder logic screen.
   c) With all switches turned off, force on pilot light 2. Use the data monitor to observe the status of the corresponding bit for pilot light 2 in the output image table file. How does forcing outputs manipulate the output image table file bits?

9. Construct the example of the subroutine instruction (SBR) and the return instruction (RET) program shown in Fig. 9-17. The purpose of the program is to find the average value of N7:5 and N7:20 and store the result in N7:30. This is accomplished by passing parameters to the subroutine and doing the math in the subroutine, and then returning the answer to the main program through the RET instruction. When S1 is closed, the data from the input parameter, N7:5, is copied into the first input parameter in the SBR instruction, N7:50. The data from the second input parameter, N7:20, is copied into the second input parameter in the SBR, N7:51. In the subroutine, N7:50 and N7:51 are then added together, with the result stored in N7:52. The value in N7:52 is then divided by 2, which gives the average of N7:50 and N7:51, with the result stored in N7:53. The RET instruction then returns the average value through the return parameter, N7:53, to N7:30 in the JSR instruction in the first rung in the main program. Prove the operation by using the data monitor to insert values for N7:5 and N7:20 and verifying that the average value is contained in N7:30.

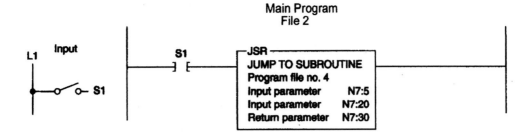

Main Program
File 2

Subroutine
File 4

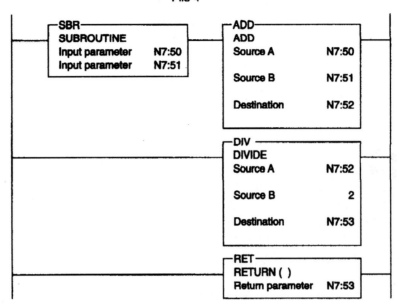

Fig. 9-17

10. Design a program to implement the flashing pilot light subroutine program described in Chapter 9 of the text (Fig. 9-10). Enter the program into the PLC and prove its operation.

11. Design a program that uses the temporary end (TND) instruction described in Chapter 9 of the text (Fig. 9-25). Demonstrate how this instruction can be used to progressively debug the program.

10 DATA MANIPULATION INSTRUCTIONS

TEST 10•1

Choose the letter that best completes the statement.

Answer

1. Data manipulation instructions enable the PLC to:
   a) move data from one memory area to another.
   b) compare data.
   c) take on some of the qualities of a computer.
   d) all of these.

1. D

2. Depending on the manufacturer, which of the following might be considered the same as a word?
   a) Register
   b) File
   c) Table
   d) All of these

2. A

3. According to the memory map of Fig. 10-1:
   a) No. 1 is a word, No. 2 is a register, No. 3 is a file.
   b) No. 1 is a register, No. 2 is a bit, No. 3 is a file.
   c) No. 1 is a file, No. 2 is a bit, No. 3 is a table.
   d) No. 1 is a bit, No. 2 is a table, No. 3 is a file.

3. B

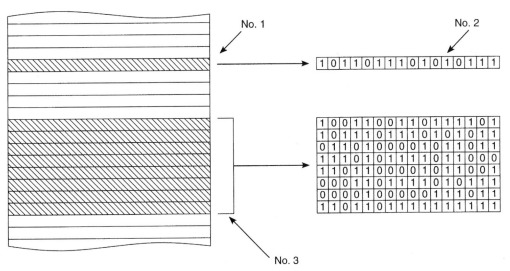

Fig. 10-1

4. The logic rung of Fig. 10-2 is an example of a:

    a) data compare instruction.

    b) data transfer instruction.

    c) timer instruction.

    d) counter instruction.

4. _B_____

5. The logic rung of Fig. 10-2 is telling the processor to get the numeric value stored in word:

    a) 320 and put it into word 500 when input *A* is true.

    b) 500 and put it into word 320 when input *A* is true.

    c) 320 and put it into word 500 when input *A* is false.

    d) 500 and put it into word 320 when input *A* is false.

5. _B_____

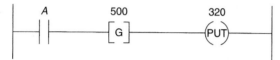

Fig. 10-2

6. In the program of Fig. 10-3, the timer starts timing when:

    a) PB1 is open.

    b) PB1 is closed.

    c) SS1 is open.

    d) SS1 is closed.

6. _B_____

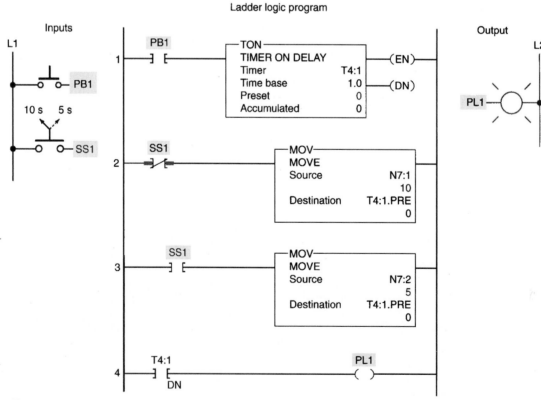

Fig. 10-3

7. In the program of Fig. 10-3, when SS1 is closed, the time-delay period will be:
   a) 000 s.
   b) 005 s.
   c) 010 s.
   d) 015 s.

8. In the program of Fig. 10-3, rung No. 3 will be true:
   a) 10 s after PB1 is closed.
   b) 5 s after SS1 is closed.
   c) 15 s after both PB1 and SS1 are closed.
   d) anytime SS1 is closed.

9. In the program of Fig. 10-3, rung No. 2 tells the processor to set the preset time of the timer to:
   a) 130 when SS1 is open.   c) 010 when SS1 is open.
   b) 130 when SS1 is closed. d) 010 when SS1 is closed.

10. In the program of Fig. 10-4, the counter increments by one for each false-to-true transition of:
   a) rung No. 1.   c) rung No. 3.
   b) rung No. 2.   d) rung No. 4.

11. In the program of Fig. 10-4, a preset count of 050 is selected when input _____ is closed.
   a) LS1   c) *B*
   b) *A*   d) *C*

12. In the program of Fig. 10-4, for the light to come on after an accumulated count of 175:
   a) rung No. 3 must be true.
   b) rungs No. 2, No. 3, and No. 4 must be true.
   c) rungs No. 1, No. 2, and No. 6 must be true.
   d) rungs No. 1, No. 2, No. 4, and No. 6 must be true.

13. Which of the following is *not* considered to be a data compare instruction?
   a) LESS THAN
   b) EQUAL
   c) GET/PUT
   d) GREATER THAN

14. Output PL1, of the logic rung shown in Fig. 10-5, will be true when the value of the number stored in word N7:10 is:
   a) less than 080.
   b) greater than 080.
   c) equal to 080.
   d) equal to or greater than 080.

Ladder logic program

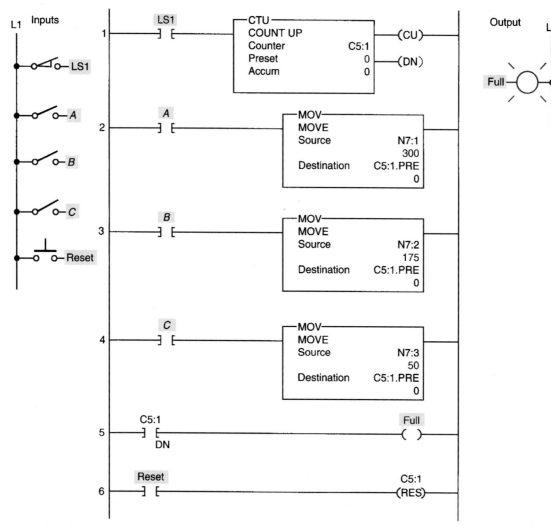

Fig. 10-4

Ladder logic program

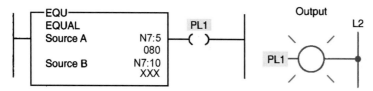

Fig. 10-5

15. The logic rung in Fig. 10-6:
   a) has logic continuity.
   b) does not have logic continuity.
   c) will cause output PL1 to be energized.
   d) both *a* and *c*.

15. _____

Ladder logic program

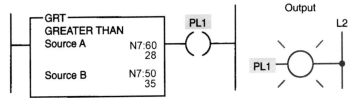

Fig. 10-6

Ladder logic program

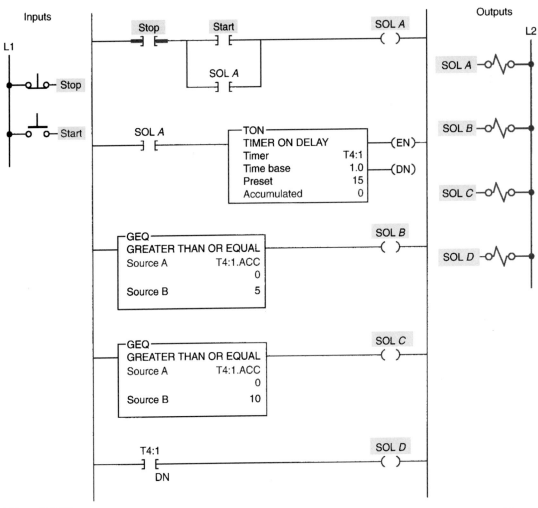

Fig. 10-7

16. In the program in Fig. 10-7, solenoid *B* is energized:

a) as soon as the start button is pressed.

b) 5 s after the start button is pressed.

c) 10 s after the start button is pressed.

d) 15 s after the start button is pressed.

16. B

17. In the program in Fig. 10-7, source *A* of the GREATER
THAN OR EQUAL instructions contains the:
a) preset time value of the timer.
b) time base value of the timer.
c) reset time value of the timer.
d) accumulated time value of the timer.

17. _____

18. In the program in Fig. 10-7, bit T4:1/DN is called the:
a) reset bit.
b) done bit.
c) timing bit.
d) data compare bit.

18. _____

19. In the program of Fig. 10-8, when switch S1 is closed:
a) the light turns on after 20 s and remains on.
b) the light turns on after 5 s and remains on.
c) the light turns on after 15 s and remains on.
d) the light turns on after 5 s, stays on for 10 s, and then turns off.

19. _____

Ladder logic program

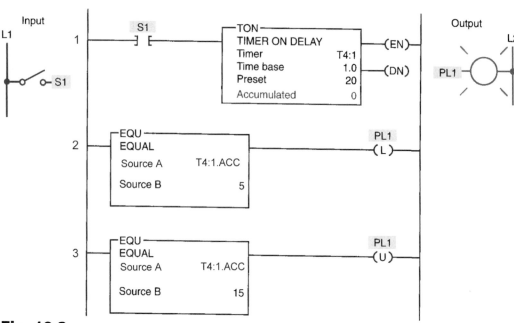

Fig. 10-8

20. In the program in Fig. 10-8, 10 s immediately after switch S1 is
closed, rung(s) _____ will be true.
a) No. 1 only
b) No. 1 and No. 2
c) No. 1 and No. 3
d) No. 1, No. 2, and No. 3

20. _____

21. In the program of Fig. 10-9, source A is addressed to the:
   a) pushbutton input.
   b) light output.
   c) done bit of the counter.
   d) accumulated value of the counter.

Ladder logic program

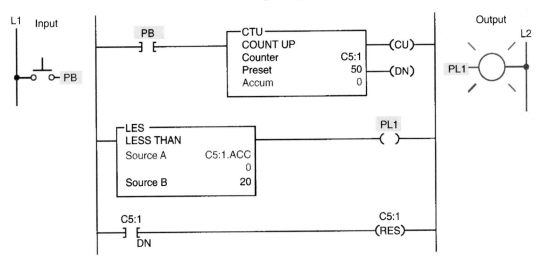

Fig. 10-9

22. In the program in Fig. 10-9, output PL1 will be true when the accumulated value of the counter is:
   a) equal to 050.
   b) equal to 020.
   c) between 000 and 019.
   d) between 020 and 050.

23. In the program in Fig. 10-9, the LESS THAN instruction is:
   a) always true.
   b) true as long as the value contained in source A is less than 20.
   c) true as long as the value contained in source A is more than 20.
   d) true as long as the value contained in source A is equal to 20.

24. In the program of Fig. 10-9, assume that the input pushbutton is pulsed 60 times after the counter had first been reset. After this operational sequence:
   a) the light would be on and the accumulated count would be 10.
   b) the light would be off and the accumulated count would be 10.
   c) the light would be on and the accumulated count would be 60.
   d) the light would be off and the accumulated count would be 60.

25. The BCD value for the thumbwheel switch setting of Fig. 10-10 would be:
   a) 1010 0110 1110 1101.
   b) 1110 0001 1010 1010.
   c) 0011 0010 1000 0100.
   d) 0010 1010 0110 1011.

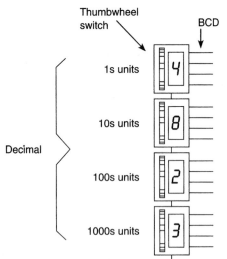

Fig. 10-10

26. A register I/O module is used to provide interface to:
  a) multibit digital devices.
  b) analog devices.
  c) discrete devices.
  d) both *a* and *b*.

26. _____

27. In an analog input module, voltages and currents are sensed and converted into digital word equivalents by a(n):
  a) D/A converter.
  b) A/D converter.
  c) rectifier.
  d) amplifier.

27. _____

28-1. For the program in Fig. 10-11, rung No. 1 contains the logic that:
  a) detects when the temperature drops below the low set point.
  b) allows the thermocouple temperature to be monitored by the LED display board.
  c) detects when the temperature rises above the high set point.
  d) switches the heater on and off.

28-1. _____

28-2. For the program in Fig. 10-11, rung No. 2 contains the logic that:
  a) detects when the temperature drops below the low set point.
  b) allows the thermocouple temperature to be monitored by the LED display board.
  c) detects when the temperature rises above the high set point.
  d) switches the heater on and off.

28-2. _____

28-3. For the program in Fig. 10-11, rung No. 3 contains the logic that:
  a) detects when the temperature drops below the low set point.
  b) allows the thermocouple temperature to be monitored by the LED display board.
  c) detects when the temperature rises above the high set point.
  d) switches the heaters on and off.

28-3. _____

28-4. For the program of Fig. 10-11, rung No. 4 contains the logic that:
a) detects when the temperature drops below the low set point.
b) allows the thermocouple temperature to be monitored by the LED display board.
c) detects when the temperature rises above the high set point.
d) switches the heaters on and off.

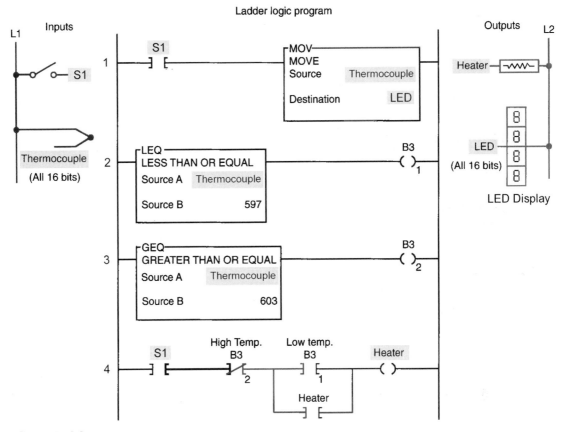

Fig. 10-11

29. In a closed-loop control system, the PLC control program acts to:
a) form a closed circuit between input and output modules.
b) monitor the output signal and adjust the input signal accordingly.
c) keep the input and output in balance.
d) correct any difference between the measured value and the desired value.

29. _____

30. The move instruction copies data from a(n) _____ word to a(n) _____ word.
a) source . . . destination
b) destination . . . source
c) integer . . . floating point
d) floating point . . . integer

30. _____

31. With the masked move instruction, where there is a _____ in the mask, data will pass.
a) 1
b) 0
c) neg
d) pos

31. _____

32. The _____ mode of the FAL instruction allows one element of data to be operated on for every false-to-true transition of the instruction.
a) all
b) numeric
c) incremental
d) sequential

32. _____

33. In a PID module the _____ mode produces an output signal that is proportional to the length of time that the error signal is present.
a) proportional
b) inverse
c) integral
d) derivative

33. _____

34. The transfer of data from a word location to a file is called a:
a) file-to-file move.
b) file-to-word move.
c) word-to-file move.
d) word-to-word move.

34. _____

TEST 10•2

Place the answers to the following questions in the answer column at the right.

Answer

1. Each data manipulation instruction requires two or more _____ of data memory for operation.

1. _____

2. The words of data memory in singular form may be referred to as either words or _____.

2. _____

3. A consecutive group of data memory words may be referred to as either a(n) (a) _____ or a(n) (b) _____.

3a. _____

3b. _____

4. The data contained in words will be in the form of binary _____ represented as series of 1's and 0's.

4. _____

5. The format used for data manipulation instructions is the same for all PLC models. (True or False)

5. _____

6. Data manipulation can be placed into the two broad categories of data (a) _____ and data (b) _____.

6a. _____

6b. _____

7. Data _____ instructions involve the transfer of the contents from one word or register to another.

7. _____

8. Data _____ instructions compare the data stored in two or more words.

8. _____

9. Data transfer instructions can address only a limited number of special locations in the memory. (True or False)

9. _____

10. GET instructions tell the processor to get a value stored in some word. (True or False)

10. _____

11. PUT instructions tell the processor where to put the information it obtained from the GET instruction. (True or False)

11. _____

12. Numerical data I/O interfaces are used to interface (a) _____ digital devices and (b) _____ devices.

12a. _____

12b. _____

13. Multibit interfaces allow a(n) _____ of bits to be input or output as a unit.

13. _____

14. Multibit interfaces are used to accommodate devices that require BCD input or outputs. (True or False)

14. _____

15. The analog input module contains a digital-to-analog converter circuit. (True or False)

15. _____

16. An analog I/O will allow monitoring and control of _____ voltages and currents.

16. _____

17. The analog output interface module receives numerical data from the processor that is translated into a proportional (a) _____ or (b) _____.

17a. _____

17b. _____

18. Set-point control in its simplest form _____ an input value to a set-point value.

18. _____

19. Four types of set-point control are: (a) _____, (b) _____, (c) _____, and (d) _____.

19a. _____

19b. _____

19c. _____

19d. _____

20. Each type of set-point control involves the use of some form of _____ control.

20. _____

21. To copy the value in N12:0 into N12:40 using the MOV instruction, you would enter: (a) _____ as the source, and (b) _____ as the destination.

21a. _____

21b. _____

22. To put the value of 0 into N10:0 through N10:150 using the FLL instruction, you would enter (a) _____ as the source, (b) _____ as the destination, and (c) _____ as the length.

22a. _____

22b. _____

22c. _____

23. To put the value in the upper half of N13:30 into the upper half of N12:0 using the MVM instruction, you would enter (a) _____ as the source, (b) _____ as the mask, and (c) _____ as the destination.

23a. _____

23b. _____

23c. _____

24. What value(s) stored in N7:2 would make this instruction true?

24. _____

```
   ┌─ EQU ────────────────┐
   │  EQUAL               │
───┤  Source A      N7:2  ├───
   │                      │
   │  Source B        25  │
   └──────────────────────┘
```

25. What value(s) stored in N7:2 would make this instruction true?

25. _____

```
   ┌─ NEQ ────────────────┐
   │  NOT EQUAL           │
───┤  Source A      N7:2  ├───
   │                      │
   │  Source B        20  │
   └──────────────────────┘
```

26. What value(s) stored in N7:2 would make this instruction true?

26. _____

```
   ┌─ LES ────────────────┐
   │  LESS THAN           │
───┤  Source A      N7:2  ├───
   │                      │
   │  Source B      N7:5  │
   │                  10  │
   └──────────────────────┘
```

27. What value(s) stored in N7:2 would make this instruction true?

27. _____

```
   ┌─ GRT ────────────────┐
   │  GREATER THAN        │
───┤  Source A      N7:2  ├───
   │                      │
   │  Source B      N7:5  │
   │                  10  │
   └──────────────────────┘
```

28. What value(s) stored in N7:2 would make this instruction true?

28. _____

```
   ┌─ GEQ ──────────────────────┐
   │  GREATER THAN OR EQUAL TO  │
───┤  Source A            N7:2  ├───
   │                            │
   │  Source B            N7:5  │
   │                        10  │
   └────────────────────────────┘
```

29. What value(s) stored in N7:2 would make this instruction true?

```
┌─ LIM ─────────────────────┐
│ LIMIT TEST                │
│ Low limit          100    │
│                           │
│ Test               N7:2   │
│                           │
│ High limit         200    │
└───────────────────────────┘
```

29. _____

30. What value(s) stored in N7:2 would make this instruction true?

```
┌─ LIM ─────────────────────┐
│ LIMIT TEST                │
│ Low limit          200    │
│                           │
│ Test               N7:2   │
│                           │
│ High limit         100    │
└───────────────────────────┘
```

30. _____

31. What value(s) stored in the source address would make this instruction true?

```
┌─ MEQ ─────────────────────┐
│ MASKED EQUAL TO           │
│ Source                    │
│                           │
│ Mask               B3:4   │
│                           │
│ Compare            B3:5   │
└───────────────────────────┘
```
B3:4 = 0000 0000 0000 1111
B3:5 = 0000 0000 0000 0111

31. _____

32. The move instruction _____ a value from one location to another.

32. _____

33. The bit distribute instruction is used to move _____ within a word or between words.

33. _____

34. Files allow large amounts of data to be scanned quickly. (True or False)

34. _____

35. For each example, signify the type of FAL copy operation shown (file-to-file, file-to-word, or word-to-file):

a)
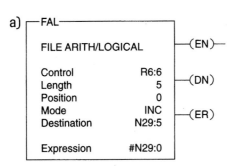

35a. _____

b)
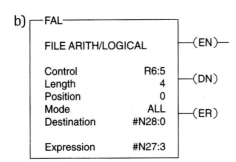

35b. _____

c)

```
┌─FAL────────────────────┐
│                         ├─(EN)─
│  FILE ARITH/LOGICAL     │
│                         │
│  Control        R6:6    ├─(DN)
│  Length            5    │
│  Position          0    │
│  Mode            ALL    ├─(ER)
│  Destination   #N29:0   │
│                         │
│  Expression     N29:5   │
└─────────────────────────┘
```

35c. _____

36. The FLL instruction shown, when true, tells the processor to (a) _____ the value of word N7:0 into the first (b) _____ words of (c) _____ N12:0.

```
┌─FLL──────────────────┐
│  FILL FILE            │
─┤                       ├─
│  Source        N7:0   │
│  Destination  #N12:0  │
│  Length           5   │
└───────────────────────┘
```

36a. _____

36b. _____

36c. _____

37. The COP instruction shown, when true, tells the processor to copy the values of the first five (a) _____ of file (b) _____ into the first five words of file (c) _____.

```
┌ COP ─────────────────────┐
│  COPY FILE               │
┤                          ├
│  Source        #N7:0     │
│  Destination   #N12:0    │
│  Length            5     │
└──────────────────────────┘
```

37a. _____

37b. _____

37c. _____

38. The COP instruction operates at a lower speed than the same operation with the FAL instruction. (True or False)

38. _____

39. The FLL instruction is frequently used to zero all the data in a file. (True or False)

39. _____

40. Data transfer and data compare instructions are both output instructions. (True or False)

40. _____

41. Input and output modules can be addressed either at the (a) _____ level or at the (b) _____ level.

41a. _____

41b. _____

42. PID control is inexpensive but not accurate enough for many applications. (True or False)

42. _____

PROGRAMMING ASSIGNMENTS FOR CHAPTER 10

1. Construct the adjustable on-delay timer data transfer program in Fig. 10-12 using any available addresses, switches, and lights on your PLC demonstration panel. After constructing your program on a separate sheet of paper, enter it into the PLC and prove its operation.

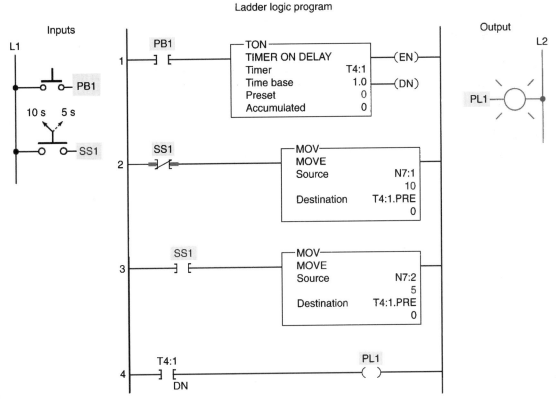

Fig. 10-12

2. Construct a simulated program for the up-counter program in Fig. 10-13. In this application, a limit switch programmed to operate a counter counts the products coming off a conveyor line onto a storage rack. Three different types of products are run on this line. The storage rack has room for only 300 boxes of product A or 175 boxes of product B or 50 boxes of product C. Three switches are provided to select the desired preset counter value, depending on whether the product line A, B, or C is being manufactured. A reset button is provided to reset the accumulated count to 0. A pilot lamp is switched on to indicate when the storage rack is full. After constructing your program on a separate sheet of paper, enter it into the PLC and prove its operation.

Ladder logic program

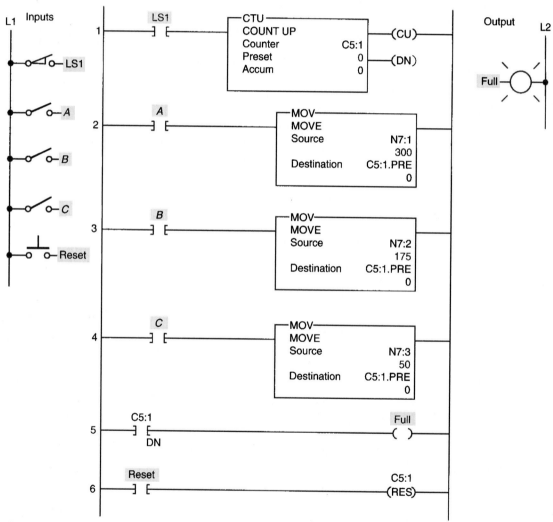

Fig. 10-13

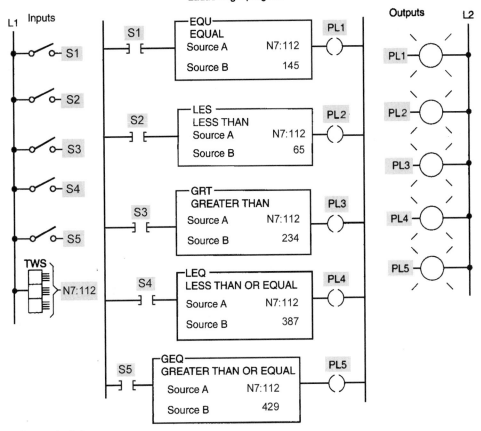

Fig. 10-14

3. Construct the data compare program in Fig. 10-14 using a thumbwheel switch interface module for the changing variable. Enter the program into the PLC and prove the operation of each rung.

4. Construct a simulated program for the *relay* time delay circuit in Fig. 10-15 using only *one* internal timer along with *data compare* statements. After constructing your program on a separate sheet of paper, enter it into the PLC and prove its operation.

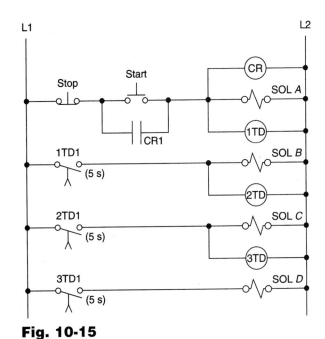

Fig. 10-15

5. Construct the on-delay timer data compare program in Fig. 10-16 using any available addresses, switches, and lights on your PLC demonstration panel. After constructing your program on a separate sheet of paper, enter it into the PLC. Demonstrate that when the switch is closed, the light comes on after 5 s, stays on for 10 s, and then turns off.

6. Construct the counter program in Fig. 10-17, using the LESS THAN instruction and any available addresses, switches, and lights on your PLC demonstration panel. After constructing your program on a separate sheet of paper, enter it into the PLC. Demonstrate that the output will be energized when the accumulated value of the counter is between 000 and 019 and that the counter will reset automatically when it reaches its preset value of 050.

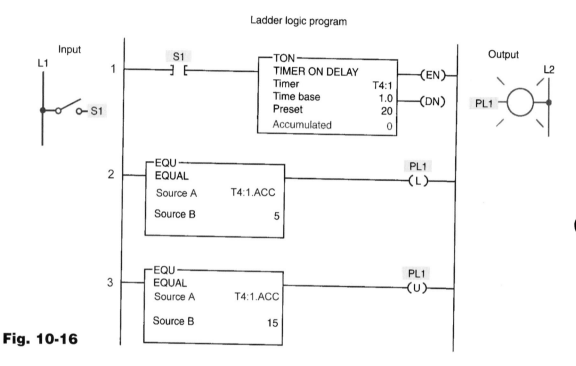

Fig. 10-16

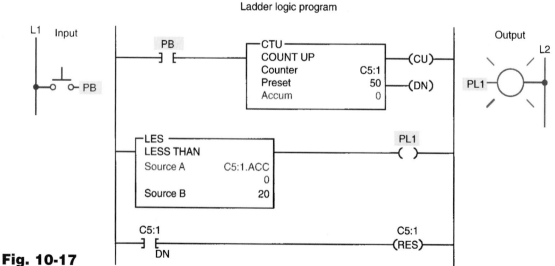

Fig. 10-17

7. Construct the counter program in Fig. 10-18 using the GREATER THAN instruction, and any available addresses, switches, and lights on your PLC demonstration panel. After constructing your program on a separate sheet of paper, enter it into the PLC. Demonstrate that the output will be energized when the counter's accumulated value is from 021 to 050 and that the counter will reset automatically when it reaches its preset value of 050.

8. Construct the BCD input/output program in Fig. 10-19 using any available addresses, switches, and lights on your PLC demonstration panel. After constructing your program on a separate sheet of paper, enter it into the PLC. Demonstrate that the decimal setting of the thumbwheel switches is monitored by the LED display board and that pilot light PL1 will turn on when switch S1 is closed and the value of the thumbwheel switches is 100.

Ladder logic program

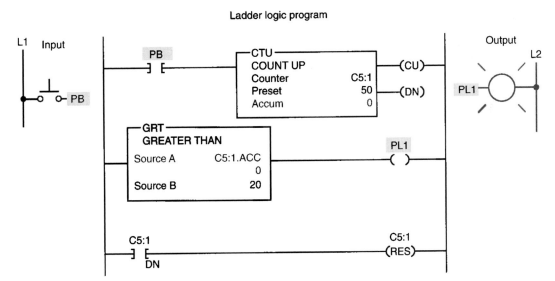

Fig. 10-18

Ladder logic program

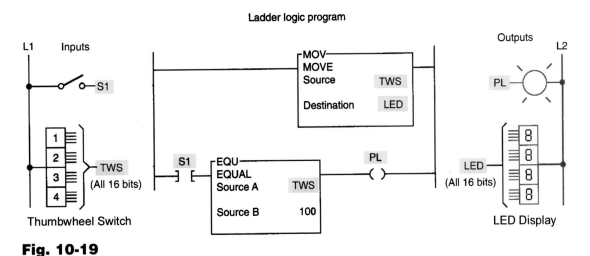

Fig. 10-19

9-1. Construct a simulated program for the temperature set-point control program in Fig. 10-20. In this application, simple off-on control of the electric heating elements of an oven is to be provided by the PLC. The oven is to maintain an average set-point temperature of 600°F, with a variation of about 1% between the off and on cycles. After constructing your program on a separate sheet of paper, enter it into the PLC and prove its operation.

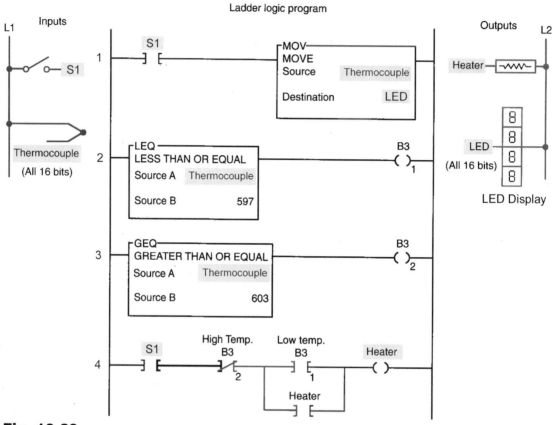

Fig. 10-20

9-2. Modify the program to include alarm lights that come on if the temperature rises above 610°F or drops below 590°F. Once on, each alarm light stays on until manually reset with a pushbutton.

10. Design a PLC program that mixes two ingredients A and B (Fig. 10-21). The mixing cycle is as follows:
   a) Ingredient A is sent to the tank first by energizing solenoid No. 1. The flow meter gives one pulse for every gallon of flow. Solenoid valve No. 1 will be open (energized) until 200 gal have poured in.
   b) After ingredient A is in the tank, 300 gal of ingredient B should be added. The process of adding follows the same procedure as ingredient A.
   c) After ingredient B is in the tank, the mixer motor starts and runs for 5 min.
   d) After the mixing is complete, solenoid No. 3 should open and let the mixed batch go into a finished tank.
   e) When the tank is empty (as indicated by the NC empty liquid-level switch), solenoid No. 3 should close and stop the cycle.

Enter the simulated program into the PLC and prove its operation.

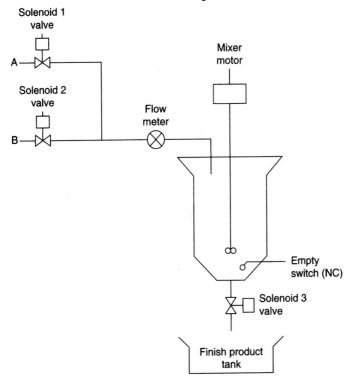

Process flow diagram

Fig. 10-21

11. Design a PLC program to control the liquid level in a tank. A level transducer is to convert the tank level to a standard process current signal of from 4 to 20 mA as illustrated in the process diagram of Fig. 10-22.

OPERATION SEQUENCE
1. The agitator motor is to be started and stopped by a pushbutton station.
2. The solenoid valves (No. 1 and No. 2) are to be open (energized) when the agitator is running.
3. Pump No. 1 starts at the 75% liquid level and stops at the 25% level. The pump will run only if the agitator is running.

4. On a high-level alarm (tank 90%), a light will come on and stay on even if the tank level drops. The operator must push a reset button to turn off the light. On a high-level alarm, the solenoids will also close (de-energize).
5. A low-level alarm light shall be provided (tank 10%), similar to the high-level, and reset using the same pushbutton.

Enter the simulated program into the PLC and prove its operation.

12-1. Modify the program of Question 11 so that, on a low-level alarm, the agitator must shut down after 30 s.

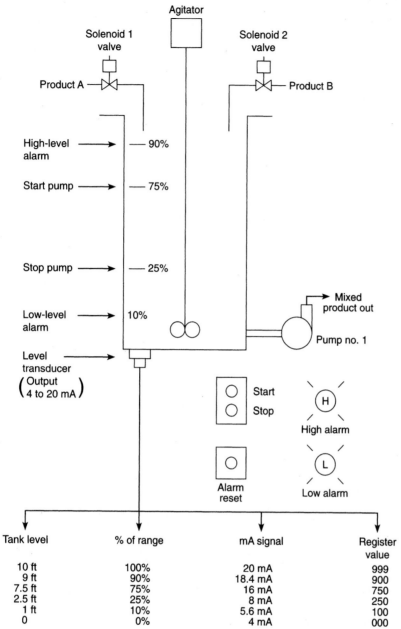

Process flow diagram

Tank level	% of range	mA signal	Register value
10 ft	100%	20 mA	999
9 ft	90%	18.4 mA	900
7.5 ft	75%	16 mA	750
2.5 ft	25%	8 mA	250
1 ft	10%	5.6 mA	100
0	0%	4 mA	000

Fig. 10-22

12-2. Enter the simulated program into the PLC and prove its operation.

13. Enter the move instruction program of Fig. 10-23 into the PLC and prove its operation.

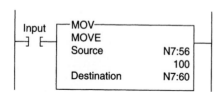

Fig. 10-23

14. Enter the masked move instruction program of Fig. 10-24 into the PLC and prove its operation.

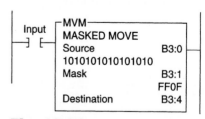

Fig. 10-24

15. Enter each of the data-comparison instructions in Fig. 10-25*a* to *i* into the PLC and prove the operation of each.

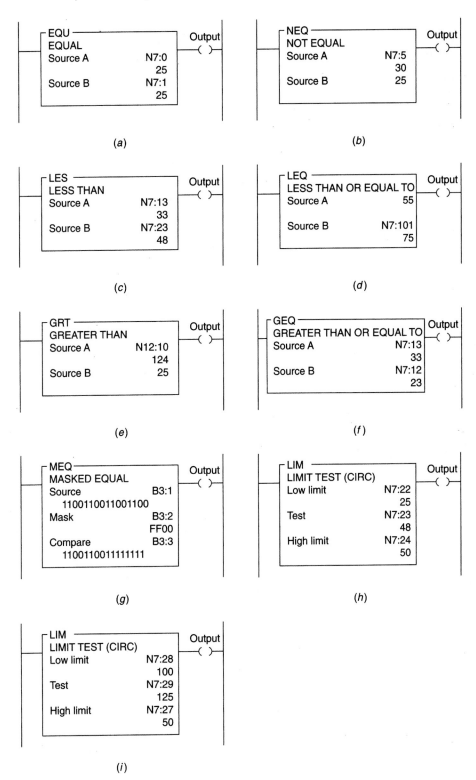

(a)

(b)

(c)

(d)

(e)

(f)

(g)

(h)

(i)

Fig. 10-25

16. Enter each of the file copy
instructions of Fig. 10-26a to e into
the PLC and prove the operation of
each.

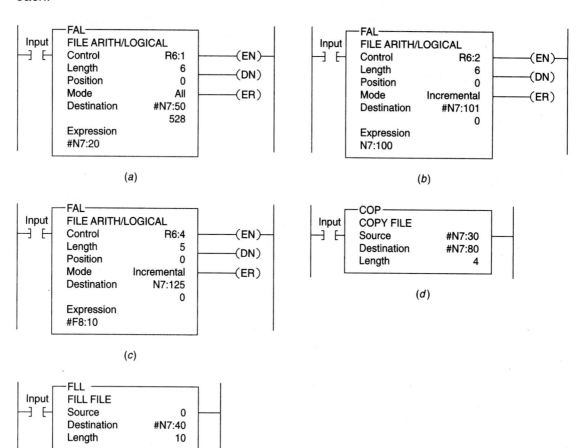

(a)

(b)

(c)

(d)

(e)

Fig. 10-26

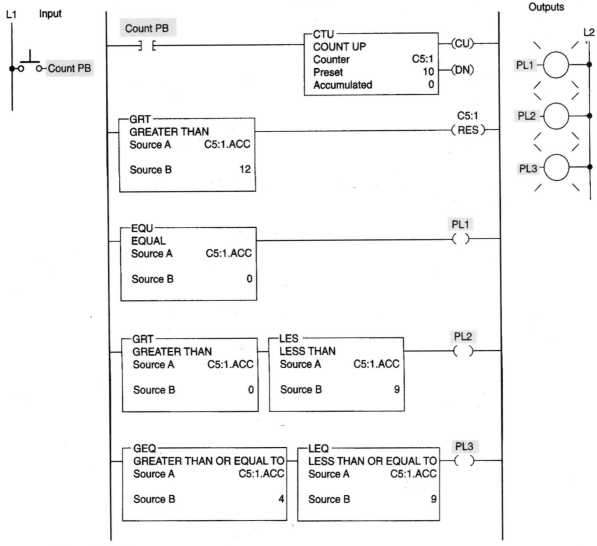

Fig. 10-27

17. Enter the data comparison program of Fig. 10-27 into the PLC. Operate the program and answer the following questions about the operation:
    a) What is the highest count achieved before the counter is reset?
    b) At what accumulated values of the counter are lights PL1, PL2, and PL3 energized?

18. Construct and test each of the following PLC data compare problems:
    a) A light is to come on only if a PLC counter has an accumulated value of 8 or 14.
    b) A light is to be on if a PLC counter does not have accumulated values of either 6 or 10.
    c) A light is to come on if three PLC counters have the same accumulated values.

19. Design and test a PLC program to implement a solution to the following problem: A room heating and air conditioning system is to be implemented with a programmable controller. The room temperature is read by a temperature transducer to word N33:1. The outdoor temperature is read by another temperature transducer to word N33:2. The logic to place these temperatures into these locations is assumed to be already in place. Design logic to:

- Turn on the heat when the indoor temperature is at or below 21°C and the outdoor temperature is below 16°C.
- Turn off the heat when the temperature is at or above 22°C.
- Turn on the air conditioning when the indoor temperature is at or above 22°C and the outdoor temperature is above 20°C.
- Turn off the air conditioning when the indoor temperature is at or below 21°C.

20. A baking process includes three ovens (No. 1, No. 2, No. 3), each controlled by a separate PLC timer. The baked product is to remain in each oven for a specified time according to the recipe produced. There are three separate recipes to run through the ovens. The following gives the bake time, in s, for each recipe:

Oven Bake Times

Recipe	No. 1	No. 2	No. 3
A	10 s	20 s	5 s
B	8 s	12 s	48 s
C	24 s	16 s	4 s

Construct a program that will allow an operator to select and run any one of the three recipes. Enter the program into the PLC and prove its operation.

MATH INSTRUCTIONS

TEST 11•1

Choose the letter that best completes the statement.

Answer

1. The ability of a PLC to perform math functions is intended to:
   a) replace a calculator.
   b) multiply the effective number of input and output devices.
   c) perform arithmetic functions on values stored in memory words.
   d) all of the above.

1. _C_

2. In the program of Fig. 11-1, the value of the number stored in N7:2 is:
   a) 172.
   b) 601.
   c) 325.
   d) 348.

2. _A_

3. In the program of Fig. 11-1, which of the following numbers stored in N7:3 will cause output PL1 to be energized?
   a) 048
   b) 124
   c) 172
   d) 325

3. _A_

Ladder logic program

```
-ADD-------------------
 ADD
 Source A        N7:0
                 124
 Source B        N7:1
                 48
 Destination     N7:2
```

```
-LES -------------------      PL1
 LESS THAN                    ( )
 Source A        N7:2
 Source B        N7:3
```

Output

PL1   L2

Fig. 11 -1

4. In the program of Fig. 11-2, assume the accumulated count
of counter C5:0 and C5:1 to be 124 and 248, respectively. As a result:
  a) the number 372 will be stored in word N7:1 and output PL1 will be
     energized.
  b) the number 372 will be stored in word N7:1 and output PL1 will not
     be energized.
  c) the number 350 will be stored in word N7:1 and output PL1 will be
     energized.
  d) the number 350 will be stored in word N7:1 and output PL1 will not
     be energized.

4. _____ A

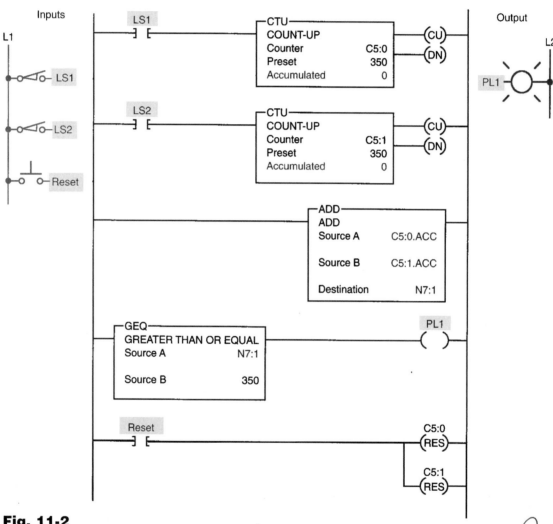

Fig. 11-2

5. In the program of Fig. 11-2, assume that the light is to come on after
a total count of 120. As a result:
  a) the preset counter C5:0 must be changed to 120.
  b) the value in source *B* of the GEQ instruction must be changed to 120.
  c) the value in source *B* of the ADD instruction must be changed to 120.
  d) the value in word N7:1 must be changed to 120.

5. _____ B

6. In the program of Fig. 11-3, the number stored in N7:2 would be:
  a) 085.
  b) 028.
  c) 181.
  d) 285.

6. _C_

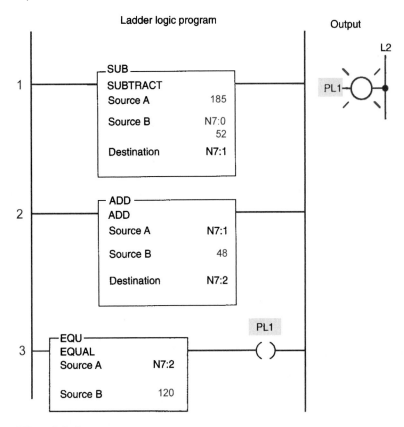

Ladder logic program   Output

Fig. 11-3

7. In the program of Fig. 11-3, rung No. 2 will be true:
  a) at all times.
  b) when the number stored in word N7:1 is equal to 048.
  c) when the number stored in word N7:1 is less than 048.
  d) when the number stored in word N7:1 is greater than 048.

7. _A_

8. In the program of Fig. 11-3, output PL1:
  a) would be energized.
  b) would not be energized.

8. _B_

9. In the program of Fig. 11-4, the preset full weight of the vessel is changed by changing:
  a) the value of the number stored at input I:012.
  b) the value of source *B* of the GEQ instruction of rung 3.
  c) the value of source *B* of the GEQ instruction of rung 5.
  d) the value of the number stored in word N7:1.

9. _B_

10. In the program of Fig. 11-4, the amount of overfill weight required to trigger the alarm is changed by changing:
   a) the value of the number stored at input I:012.
   b) the value of source *B* of the GEQ instruction of rung 3.
   c) the value of source *B* of the GEQ instruction of rung 5.
   d) the value of the number stored in word N7:1.

10.

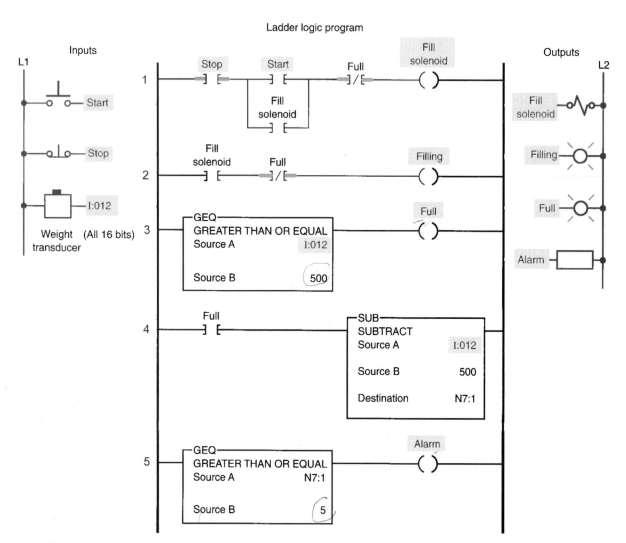

Fig. 11-4

11. In the program of Fig. 11-4, when the Full light is on:
   a) the weight of the vessel is 500 pounds or more.
   b) rung No. 3 is always true.
   c) rung No. 1 is always false.
   d) all of these.

11.

12. In the program of Fig. 11-4, when the Filling light is on:
   a) the weight of the vessel is less than 500 pounds.
   b) rung No. 2 is always true.
   c) rung No. 4 is always false.
   d) all of these.

12.

13. In the program of Fig. 11-4, the number stored in word N7:1 represents the:

a) weight of the empty vessel.
b) preset weight of the vessel.
c) current weight of the vessel.
d) difference between the current and preset weight of the vessel.

13.

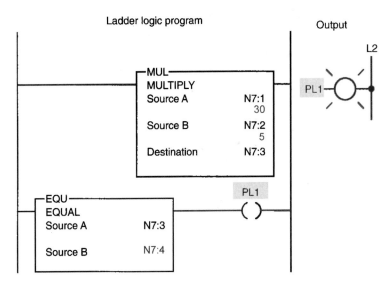

Fig. 11-5

14. In the program of Fig. 11-5, the number stored in N7:3 is:

a) 6.
b) 60.
c) 150.
d) 300.

14. _C_

15. In the program of Fig. 11-5, the number stored in N7:3 will turn PL1 on?

a) 150
b) 100
c) 50
d) all of these

15.

16. In the program of Fig. 11-5, assume the value stored in N7:1 changes from 30 to 10. Which value stored in N7:4 will result in PL1 being energized?

a) 10
b) 25
c) 35
d) 50

16. _D_

17. In the program of Fig. 11-6, the set-point temperature is set by the number stored in:

a) N7:0.
b) N7:2.
c) I:012.
d) I:013.

17. _____

18. In the program of Fig. 11-6, the number stored in N7:1 represents the: **18.** _____
   a) upper temperature limit.
   b) lower temperature limit.
   c) current temperature of the oven.
   d) difference between the preset and current temperature.

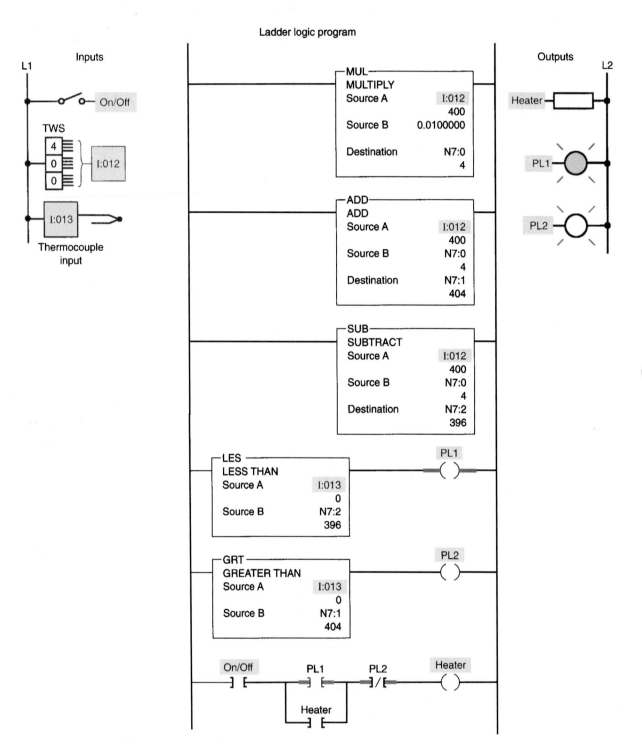

Fig. 11-6

19. In the program of Fig. 11-6, PL1 will be on whenever the current temperature is:
a) greater than the preset temperature.
b) less than the preset temperature.
c) greater than the upper temperature limit.
d) less than the lower temperature limit.

19. _____

20. In the program of Fig. 11-6, the ADD instruction is telling the processor to add the:
a) preset and current temperatures.
b) upper and lower temperature limits.
c) current and upper limit temperatures.
d) preset and upper deadband range.

20. _____

21. In the program of Fig. 11-6, assume the set-point temperature is changed to 200°F. As a result, the number stored in N7:0 would be:
a) 2.
b) 4.
c) 6.
d) 8.

21. _____

22. In the program of Fig. 11-6, assume the upper and lower temperature limits are programmed for 2% instead of 1% and the preset is 400°F. As a result, the number stored in N7:2 would be:
a) 392.
b) 390.
c) 388.
d) 386.

22. _____

23. In the program of Fig. 11-7, the number stored in N7:5 would be:
a) 1000.
b) 500.
c) 4.
d) 2.

23. _____

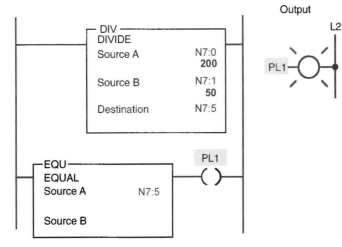

Fig. 11-7

24. In the program of Fig. 11-7, what constant stored in source *B* of the EQU instruction would turn PL1 on?
a) 24
b) 20
c) 6
d) 4

24. _____

25. In the program of Fig. 11-7, assume the value stored in N7:0 is 90, the value stored at N7:1 is 3 and the constant for source *B* of the EQU instruction is 10. What would the state of PL1 be?
a) off
b) on

25. _____

26. The program of Fig. 11-8 is used to convert the Celsius temperature indicated by the thumbwheel switch to Fahrenheit values for display. Answer each of the following with reference to this program, assuming a thumbwheel switch setting of 25°C.

Ladder logic program

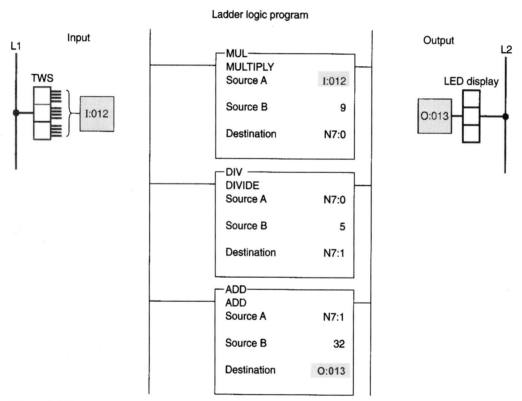

Fig. 11-8

26-1. The value of the number stored in I:012 is:
a) 25.
b) 30.
c) 35.
d) 40.

26-1. _____

26-2. The value of the number stored in N7:0 is:

a) 225.

b) 500.

c) 750.

d) 230.

26-2. _____

26-3. The value of the number stored in N7:1 is:

a) 90.

b) 45.

c) 35.

d) 60.

26-3. _____

26-4. The value of the number stored in O:013 is:

a) 98.

b) 77.

c) 67.

d) 57.

26-4. _____

27. Math instructions are all _____ instructions.

a) output

b) input

c) binary

d) BCD

27. _____

28. File arithmetic functions are used to perform arithmetic operations on:

a) multiple words.

b) integer numbers only.

c) decimal numbers only.

d) BCD numbers only.

28. _____

29. In the instruction of Fig. 11-9, the _____ used as part of the expression is a constant.

a) Source A

b) 0

c) 300

d) N7:20

29. _____

```
┌─ADD ─────────────────┐
│ ADD                  │
│ Source A      N7:15  │
│                  0   │
│ Source B       300   │
│                      │
│ Destination   N7:20  │
│                  0   │
└──────────────────────┘
```

Fig. 11 -9

30. Which math instruction would you use if you wanted to take the opposite sign of a value?

a) SUB

b) SQR

c) NEG

d) CLR

30. _____

31. Which math instruction would you use if you wanted to calculate the difference between the accumulated values of two counters?
a) ADD
b) SUB
c) MUL
d) DIV

31. _____

32. With reference to Fig. 11-10, the value of the number stored at Source B is:
a) N7:8
b) N7:16
c) 328
d) 528

32. _____

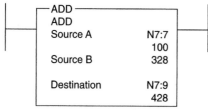

Fig. 11-10

33. With reference to Fig. 11-11, the value of the number stored at Destination is:
a) 293
b) −193
c) 51
d) −342

33. _____

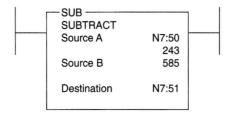

Fig. 11-11

34. With reference to Fig. 11-12, the value of the number stored at N7:30 is:

    a) 5
    b) 15.5
    c) 4.87
    d) 1.85

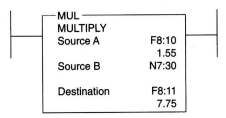

Fig. 11-12

35. With reference to Fig. 11-13, the value of the number stored at Destination is:

    a) 17
    b) 16.666
    c) 50
    d) 51

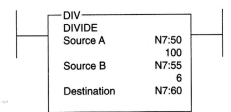

Fig. 11-13

TEST 11•2

Place the answers to the following questions in the answer column at the right.

Answer

1. Math instructions enable the programmable controller to take on some of the qualities of a(n) _____ system.

1. _____

2. The ability of a PLC to perform math functions is intended to allow it to replace a calculator. (True or False)

2. _____

3. PLC math functions perform arithmetic on _____ stored in memory words.

3. _____

4. The four basic math functions performed by PLCs are: (a) _____, (b) _____, (c) _____, and (d) _____.

4a. _____

4b. _____

4c. _____

4d. _____

5. All PLC manufacturers use the same format for math instructions. (True or False)

5. _____

6. The rung in Fig. 11-14 is telling the processor to add the values stored in words (a) _____ and store the (b) _____ in word (c) _____ whenever (d) _____ is true.

6a. _____

6b. _____

6c. _____

6d. _____

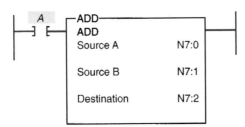

Fig. 11-14

7. The rung in Fig. 11-15 is telling the processor to subtract the value stored in word (a) _____ from the value stored in word (b) _____ and store the (c) _____ in word (d) _____.

7a. _____

7b. _____

7c. _____

7d. _____

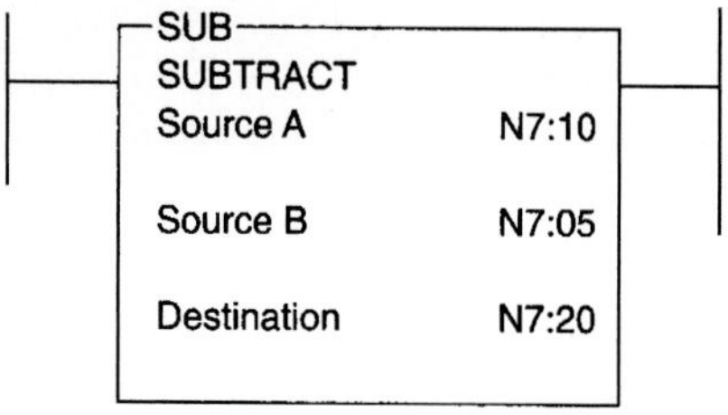

```
┌─SUB──────────────────┐
│ SUBTRACT             │
│ Source A      N7:10  │
│                      │
│ Source B      N7:05  │
│                      │
│ Destination   N7:20  │
└──────────────────────┘
```

Fig. 11-15

8. The rung in Fig. 11-16 is telling the processor to multiply the values stored in words (a) _____ and store the (b) _____ in word (c) _____.

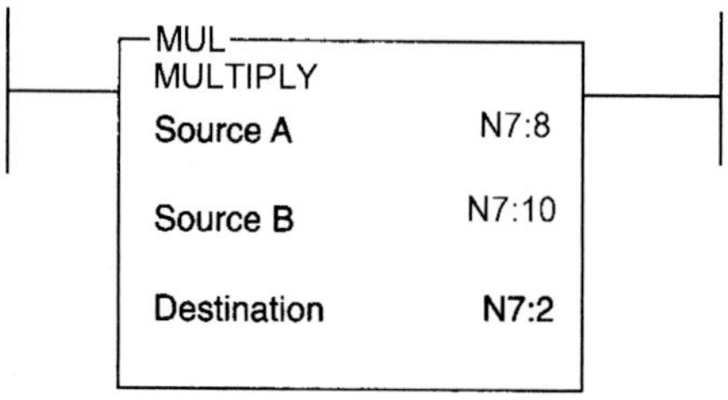

8a. _____

8b. _____

8c. _____

```
┌─MUL──────────────────┐
│ MULTIPLY             │
│ Source A      N7:8   │
│                      │
│ Source B      N7:10  │
│                      │
│ Destination   N7:2   │
└──────────────────────┘
```

Fig. 11-16

9. The rung in Fig. 11-17 is telling the processor to divide the value of word (a) _____ by the value of word (b) _____ and store the (c) _____ in word (d) _____.

9a. _____

9b. _____

9c. _____

9d. _____

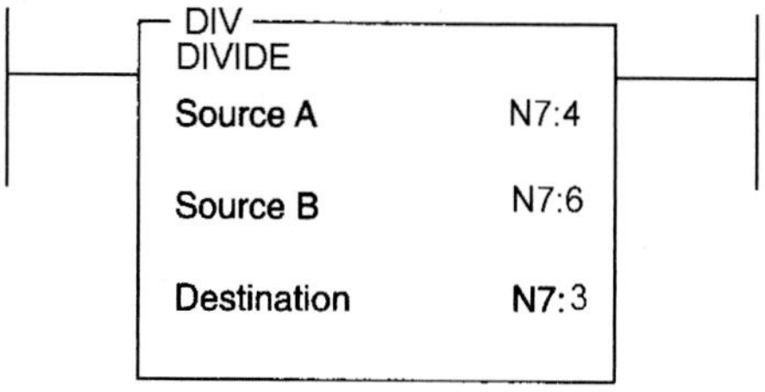

```
┌─DIV──────────────────┐
│ DIVIDE               │
│ Source A      N7:4   │
│                      │
│ Source B      N7:6   │
│                      │
│ Destination   N7:3   │
└──────────────────────┘
```

Fig. 11-17

10. The program in Fig. 11-18 is telling the processor to energize output O:2/3 whenever the sum of the values stored in words (a) _____ is (b) _____ or (c) _____ the value stored in word (d) _____.

10a. _____

10b. _____

10c. _____

10d. _____

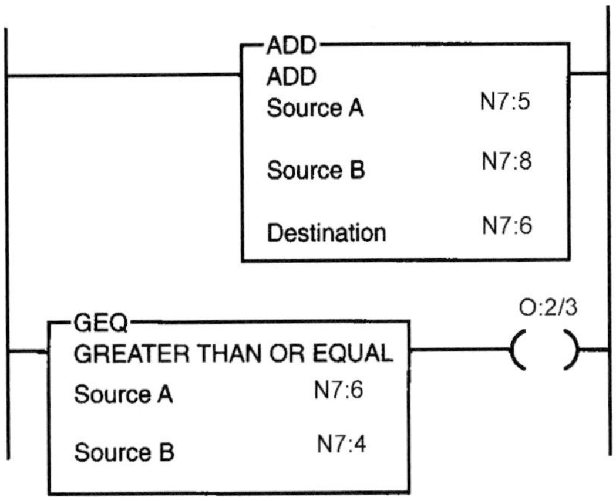

```
             ┌─ADD──────────────────┐
             │ ADD                   │
─────────────┤ Source A      N7:5    ├──────────
             │                       │
             │ Source B      N7:8    │
             │                       │
             │ Destination   N7:6    │
             └───────────────────────┘

      ┌─GEQ─────────────────────┐        O:2/3
      │ GREATER THAN OR EQUAL    │       ─( )─
──────┤ Source A      N7:6       ├────────
      │                          │
      │ Source B      N7:4       │
      └──────────────────────────┘
```

Fig. 11-18

11. The program in Fig. 11-19 is telling the processor to energize output O:6/2 whenever (a) _____ is true and the difference between the values stored in words (b) _____ is equal to the value stored in word (c) _____.

11a. _____

11b. _____

11c. _____

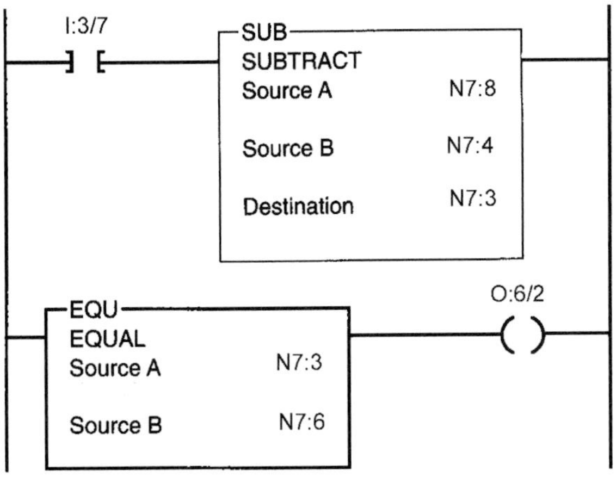

```
  I:3/7      ┌─SUB──────────────────┐
──┤ ├────────┤ SUBTRACT              ├──────────
             │ Source A      N7:8    │
             │                       │
             │ Source B      N7:4    │
             │                       │
             │ Destination   N7:3    │
             └───────────────────────┘

      ┌─EQU─────────────────┐            O:6/2
      │ EQUAL                │          ─( )─
──────┤ Source A      N7:3   ├──────────
      │                      │
      │ Source B      N7:6   │
      └──────────────────────┘
```

Fig. 11-19

12. In the program in Fig. 11-20, if output O:5/8 is to be energized when the product of the values stored in words N7:1 and N7:2 is equal to 1520, then the value of the number stored in word N7:4 must be _____.

12. _____

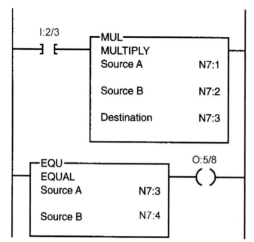

Fig. 11-20

13. In the program in Fig. 11-21, assume output O:3/2 is energized and the values of the numbers stored in words N7:0 and N7:1 are 500 and 40, respectively. The value of the number stored in word N7:8 would be _____.

13. _____

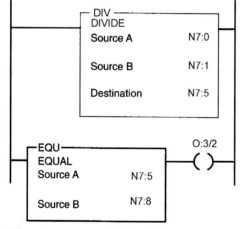

Fig. 11-21

14. In the program in Fig. 11-22, assume the value of the numbers stored in words N7:1, N7:2, and N7:6 are 600, 750, and 100, respectively. As a result, output O:3/5 will be (a) _____, the number stored in word N7:3 will be (b) _____, and output O:4/7 will be (c) _____.

14a. _____

14b. _____

14c. _____

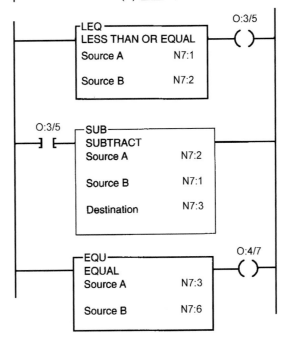

Fig. 11-22

15. In the program in Fig. 11-23, assume the value of the numbers stored in words N7:1, N7:2, N7:4, and N7:6, are 40, 9, 5, and 32, respectively. As a result, the value of the number stored in N7:3 is (a) _____, in N7:5 is (b) _____, and in N7:8 is (c) _____.

15a. _____

15b. _____

15c. _____

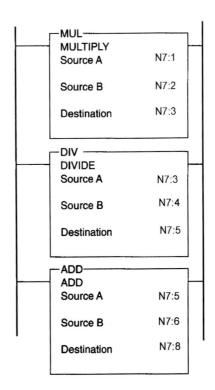

Fig. 11-23

16. Math instructions are all output instructions. (True or False)

16. _____

17. There is no limit to the maximum value a PLC math function can store. (True or False)

17. _____

18. File arithmetic instructions are designed to perform math operations on single words. (True or False)

18. _____

19. Identify the math function for each of the following FAL instructions:

19a. _____

19b. _____

a)
```
┌ FAL ─────────────────────────┐
│ FILE ARITH/LOGICAL           │
│ Control            R6:1      │
│ Length                6      │
│ Position              0      │
│ Mode                All      │
│ Destination      #F8:200     │
│                              │
│ Expression                   │
│ SQR #F8:100                  │
└──────────────────────────────┘
```

c)
```
┌ FAL ─────────────────────────┐
│ FILE ARITH/LOGICAL           │
│ Control            R6:8      │
│ Length                4      │
│ Position              0      │
│ Mode                All      │
│ Destination      #N7:500     │
│                              │
│ Expression                   │
│ #N7:330 * N7:23              │
└──────────────────────────────┘
```

19c. _____

19d. _____

19e. _____

b)
```
┌ FAL ─────────────────────────┐
│ FILE ARITH/LOGICAL           │
│ Control            R6:5      │
│ Length                4      │
│ Position              0      │
│ Mode                  2      │
│ Destination      #N7:255     │
│                              │
│ Expression                   │
│ #N10:0 – 255                 │
└──────────────────────────────┘
```

d)
```
┌ FAL ─────────────────────────┐
│ FILE ARITH/LOGICAL           │
│ Control            R6:1      │
│ Length                4      │
│ Position              0      │
│ Mode                All      │
│ Destination      #N7:100     │
│                              │
│ Expression                   │
│ #N7:25 + #N7:50              │
└──────────────────────────────┘
```

e)
```
┌ FAL ─────────────────────────┐
│ FILE ARITH/LOGICAL           │
│ Control            R6:7      │
│ Length                4      │
│ Position              1      │
│ Mode         Incremental     │
│ Destination        F8:200    │
│                       0.1    │
│ Expression                   │
│ #F8:20 I #F8:100             │
└──────────────────────────────┘
```

20. With reference to Fig. 11-24, when the input goes true, determine the value that will be stored in each of the following words:

a) N7:3
b) N7:5
c) F8:1

20a. ______________________

20b. ______________________

20c. ______________________

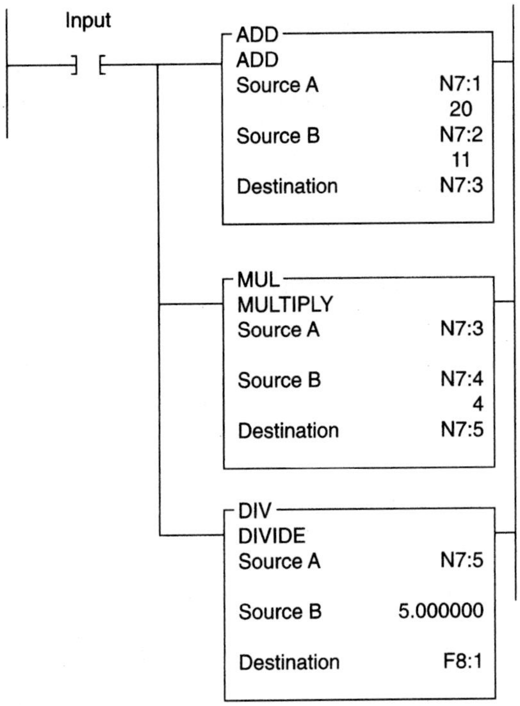
Input
ADD
ADD
Source A N7:1
20
Source B N7:2
11
Destination N7:3
MUL
MULTIPLY
Source A N7:3
Source B N7:4
4
Destination N7:5
DIV
DIVIDE
Source A N7:5
Source B 5.000000
Destination F8:1

Fig. 11-24

21. With reference to Fig. 11-25, when the input goes true, determine the value that will be stored in each of the following words:

   a) N7:0

   b) N7:1

   c) N7:2

   d) N7:3

   e) N7:4

   f) F8:0

21a. _____

21b. _____

21c. _____

21d. _____

21e. _____

21f. _____

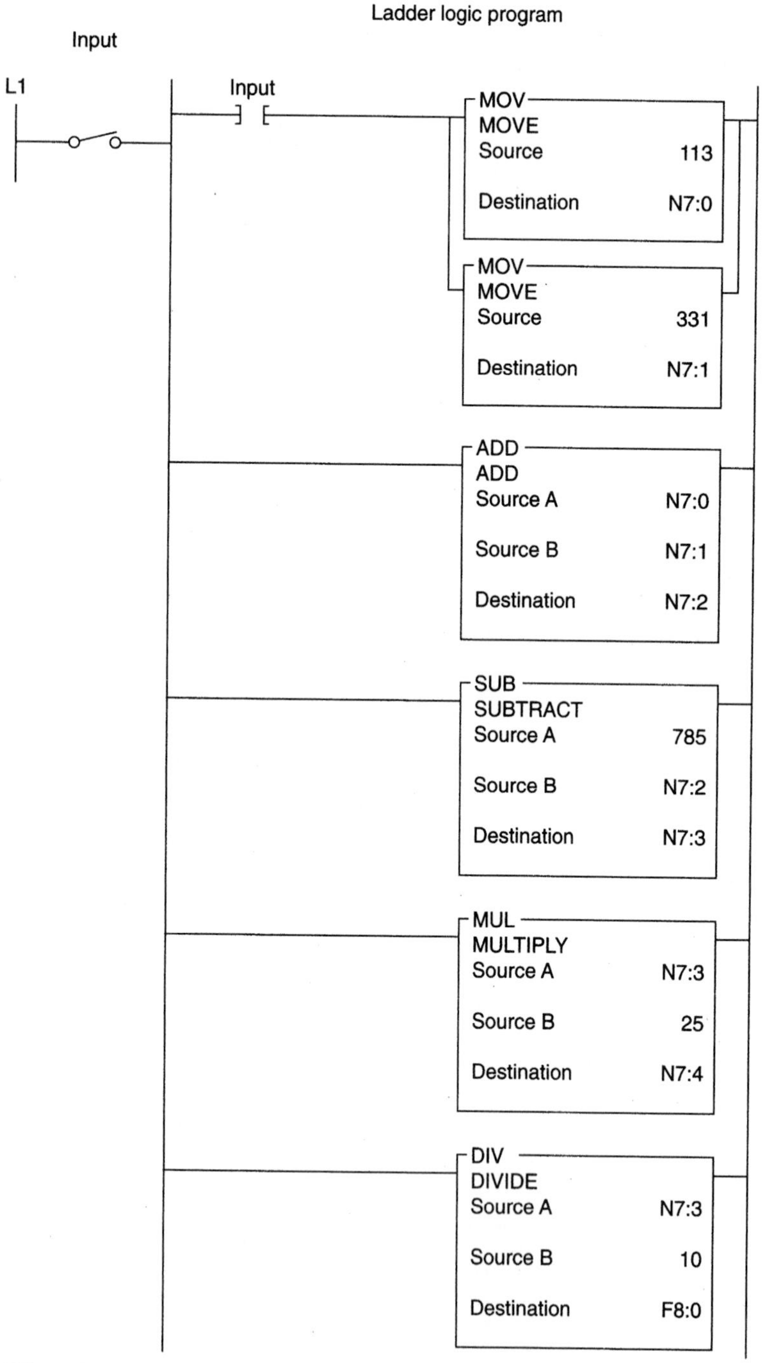

Ladder logic program

Fig. 11-25

22. With reference to Fig. 11-26, when the input goes true, determine the value that will be stored in each of the following words:
a) N7:100
b) N7:101
c) N7:102
d) N7:103

22a. _____

22b. _____

22c. _____

22d. _____

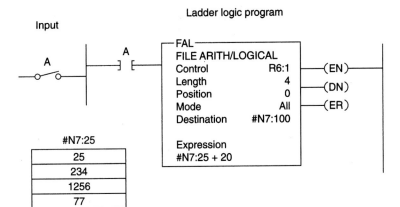

Fig. 11-26

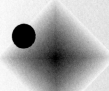

PROGRAMMING ASSIGNMENTS FOR CHAPTER 11

1. Construct a simulated program for the counter program in Fig. 11-27 using any available addresses, switches, and lights on your PLC demonstration panel. This application requires a light to come on when the sum of the counts of the two counters is equal to or greater than 350. After constructing your program on a separate sheet of paper, enter the program into the PLC and prove its operation.

Ladder logic program

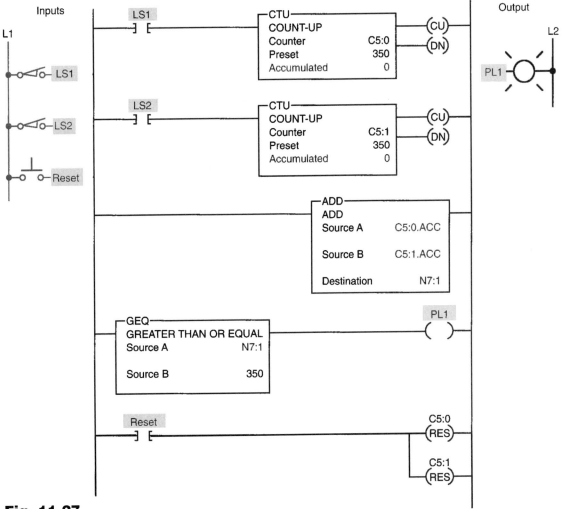

Fig. 11-27

2. Modify the program of Question 1 so that a second light comes on when the accumulated count of the two counters is equal to 345 and remains on until the reset button is pressed. Enter the modified program into the PLC and prove its operation.

3. Construct a simulated program for the overfill alarm program in Fig. 11-28. This application requires an alarm to sound when a supply system leaks 5 lbs or more of raw material into the vessel after a preset weight of 500 lbs has been reached. After constructing your program on a separate sheet of paper, enter the program into the PLC and prove its operation.

4. Modify the program of Question 3 so that, should an overfill condition of 5 lbs or more occur, an overfill solenoid is energized to automatically reduce the level back down to the 500 lbs point. Enter the modified program into the PLC and prove its operation.

5. Construct a simulated program for the oven temperature control program in Fig. 11-29. In this application, the PLC calculates the upper and lower *deadband* or on/off limits about the setpoint. The upper and lower limits are set automatically at ±1% regardless of the set-point value. The set-point temperature is adjustable by

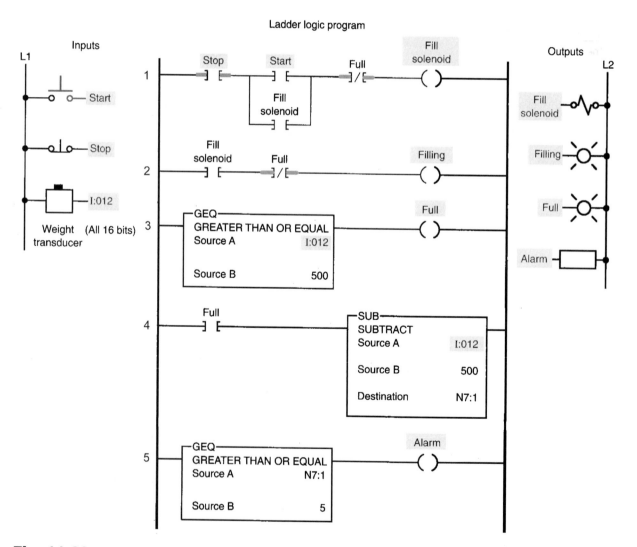

Fig. 11-28

means of thumbwheel switches, and an analog thermocouple interface module is used to monitor the current temperature of the oven. In this example, the set-point temperature is 400°F. Therefore, the electric heaters will be turned on when the temperature of the oven drops to less than 396°F and stay on until the temperature rises above 404°F. After constructing your program on a separate sheet of paper, enter it into the PLC and prove its operation.

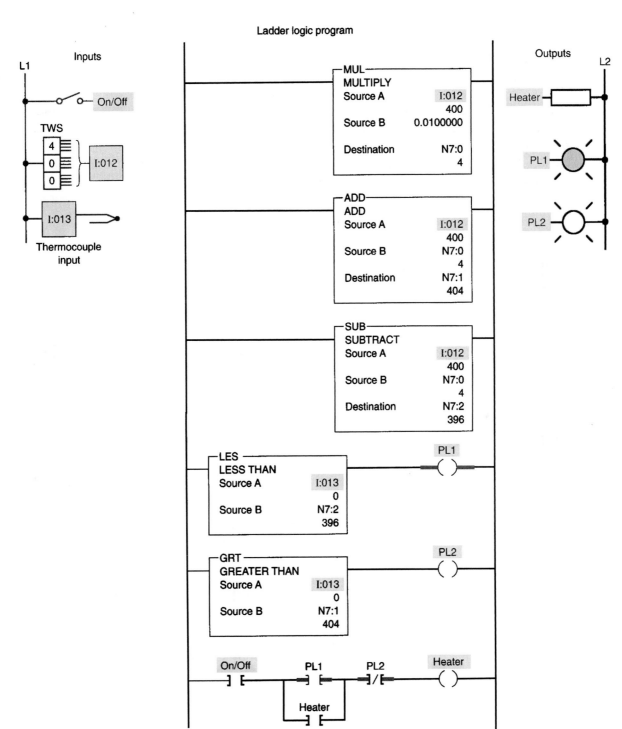

Fig. 11-29

6. Modify the program of Question 5 to include each of the following:
   a) An LED output module to display the actual temperature
   b) A high temperature light to come on if the temperature rises above 410°F
   c) A low temperature light to come on if the temperature drops below 390°F

Enter the modified program into the PLC and prove its operation.

7. Construct a program for the Celsius to Fahrenheit conversion program in Fig. 11-30. In this application, the thumbwheel switch connected to the input module indicates Celsius temperature. The program is designed to convert the recorded Celsius temperature in the data table to Fahrenheit values for display. After constructing your program on a separate sheet of paper, enter it into the PLC and prove its operation.

8. Design a simulated PLC program that will control the temperature of a furnace and monitor the temperature between 87°C and 100°C. An analog thermocouple input that measures Celsius temperature is to be used.

OPERATIONAL SEQUENCE
   1. The sensed Celsius temperature is to be converted to Fahrenheit for display.
   2. When the displayed temperature drops below 190°F for a minimum of 5 s, a heater is turned on to bring the temperature back into the desired range. The heater stays on until the temperature rises back to 190°F.
   3. Should the displayed temperature reach 212°F, an alarm is turned on and remains on until manually reset with a pushbutton.

After constructing your program on a separate sheet of paper, enter it into the PLC and prove its operation.

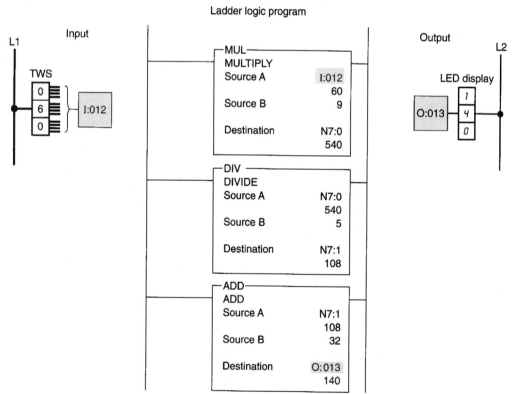

Ladder logic program

Fig. 11-30

9. Enter each of the following math operations (a through j) into the PLC and prove the operation of each.

a)
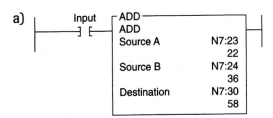

b)
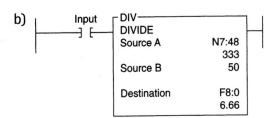

c)
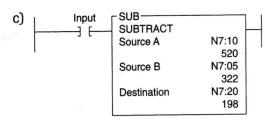

d)
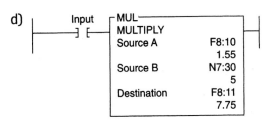

e)
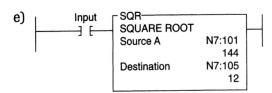

f)
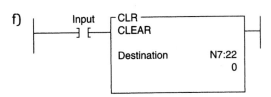

g)
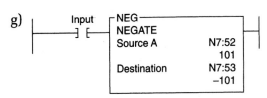

h)
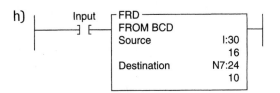

i)
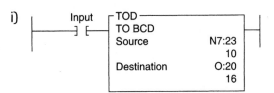

j)
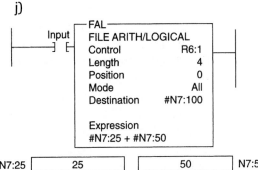

N7:25	25		50	N7:50
	234		22	
	1256		456	
N7:28	77		100	N7:53

10. Design a program that will implement the following arithmetic operation:
- Using a move instruction, place the value of 16 in N7:1 and 48 in N7:2.
- Add the values together and store the result in N7:3.
- Subtract the value in N7:3 from 650 and store the result in N7:4.
- Multiply the value in N7:4 by 15 and store the result in N7:5.
- Divide the value in N7:4 by 18 and store the result in F8:1.

Enter the program into the PLC and prove its operation.

11. Construct a program that will determine the average value of the accumulated value from four counters. Enter the program into the PLC and prove its operation.

12. A conveyor has 6-, 8-, and 12-packs of canned soda entering it. Each size of entering pack has an individual pack quantity counter. To know how many cans enter the conveyor, set up a program for multiplying and then adding to give a total can count. Enter the program into the PLC and prove its operation.

13. Write a program that will implement the following arithmetic operation:
  - Using a move instruction, place the value 30 in N7:1 and 25 in N7:2.
  - Multiply the values together and store the result in N7:3.
  - Add the value 115 to the value stored in N7:3.
  - Subtract the value 325 from the value stored in N7:3 and store the result in N7:4.
  - Divide the value in N7:3 by 5 and store the result in F8:1.

Enter the program into the PLC and prove its operation.

14. a) Use a file arithmetic logic (FAL) instruction to copy a table or file of data from N7:0–4 to N7:5–9. Add the two files together and store the results at N7:10–14.
  b) Repeat part *a* using subtraction, multiplication, division, and square root expressions.

15. Two parts-conveyor lines, A and B, feed a main conveyor line M. A third conveyor line, R, removes rejected parts a short distance down from the main conveyor. Conveyors A, B, and R have parts counters connected to them. Construct a PLC program to obtain the total parts output of main conveyor M. Enter the program into the PLC and prove its operation.

12

SEQUENCER AND SHIFT REGISTER INSTRUCTIONS

TEST 12•1

Choose the letter that best completes the statement.

Answer

1. Which of the following would *not* be classified as a sequencer switch?
   a) Rotary switch
   b) Pressure switch
   c) Drum switch
   d) Stepper switch

1. B

2. The equivalent data table for the sequencer drawn in Fig. 12-1 would be:
   a) 0000 0000
      1111 1111
      0101 0101
      1010 1010
      1111 0000
   b) 1101 0110
      0110 0101
      0100 0100
      1010 1010
      0010 1011
   c) 0010 1011 1
      1010 1010 2
      0100 0100 3
      1100 1100 4
   d) 0110 1011
      1010 1100
      1001 1101
      1001 1111
      0001 1010

2. C

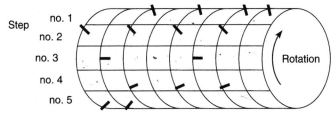

Fig. 12-1

3. Sequencer switches are used whenever:
   a) a counter function is required.
   b) a timer function is required.
   c) a time-delay function is required.
   d) a repeatable operating pattern is required.

3. _____ D

4. In the sequencer circuit of Fig. 12-2, the total length of time (assuming 45 s per increment) that the heater is on for one complete cycle is approximately:
   a) 35 min.
   b) 40 min.
   c) 20 min.
   d) 56 min.

4. _____ A

5. In the sequencer circuit of Fig. 12-2, when the timer is at the 14-min point in the cycle, only the _____ outputs will be on.
   a) timer and heater
   b) timer, drain pump, and circulating motor
   c) timer, heater, and circulating motor
   d) timer, fill value, heater, and circulating motor

5. _____ D

6. If the sequencer circuit of Fig. 12-2 were to be implemented using a PLC, which of the following output devices would *not* be required?
   a) Heater
   b) Timer
   c) Drain pump
   d) Safety water level switch

6. _____ B

7. If the sequencer circuit of Fig. 12-2 were to be implemented using a PLC, the number of programmed real outputs required would be:
   a) 4.
   b) 5.
   c) 6.
   d) 7.

7. _____ A

8. The information for each PLC sequencer step is entered into:
   a) the output module.
   b) the input module.
   c) the programmer.
   d) a word file.

8. _____ D

9. As the PLC sequencer advances through its steps, information is transferred from:
   a) the output module to the input module.
   b) the input module to the output module.
   c) the programmer to the processor.
   d) the word file to the output word.

9. _____ D

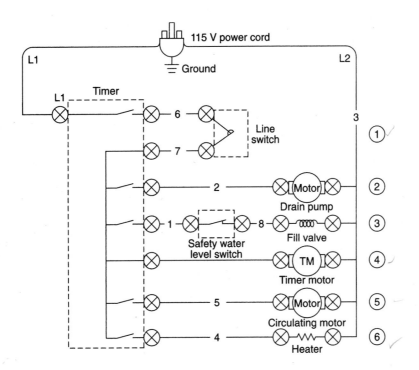

Fig. 12-2

Machine function		Timer increment	Active circuits			
Off		0–1				
First prerinse	Drain	2	1 2	4		
	Fill	3	1	3 4 5		
	Rinse	4–5	1	4 5 6		
	Drain	6	1 2	4 5		
Prewash	Fill	7	1	3 4 5		
	Wash	8–10	1	4 5 6		
	Drain	11	1 2	4 5		
Second prerinse	Fill	12	1	3 4 5		
	Rinse	13–15	1	4 5 6		
	Drain	16	1 2	4		
Wash	Fill	17	1	3 4		
	Wash	18–30	1	4 5 6		
	Drain	31	1 2	4 5		
First rinse	Fill	32	1	3 4 5		
	Rinse	33–34	1	4 5 6		
	Drain	35	1 2	4 5		
Second rinse	Fill	36	1	3 4 5		
	Rinse	37–41	1	4 5 6		
	Drain	42	1 2	4 5		
Dry	Dry	43–58	1	4 6		
	Drain	59	1 2	4 6		
	Dry	60	1	4 6		

10. On some PLCs, a mask word is used to:
a) selectively screen out data.
b) reset the sequencer automatically after the last step.
c) advance the sequencer to the next step.
d) all of these.

10. ____

11. With reference to the PLC sequencer instruction of Fig. 12-3, words 121, 122, 123, and 124 make up a:
a) five-word file.
b) four-word file.
c) five-word output.
d) four-word output.

11. ___B___

12. With reference to the PLC sequencer instruction of Fig. 12-3, bit 7 of output word 100 will be:
a) 0 for each step.
b) 1 for each step.
c) 1 for steps 1 and 4 only.
d) 1 for steps 2 and 3 only.

12. ___C___

13. With reference to the PLC sequencer instruction of Fig. 12-3, if word 123 is to be entered using the hexadecimal code, it would be entered as:
a) 64AB.
b) 12CD.
c) 46EF.
d) 20BA.

13. ___A___

14. With reference to the PLC sequencer instruction of Fig. 12-3, it would be possible to mask bit _____ of output word 100.
a) 5
b) 9
c) 13
d) all of these.

14. ___D___

15. Typically, the sequencer program can do in 20 memory words what a standard program can do in:
a) 5 words.
b) 10 words.
c) 20 words.
d) 100 words.

15. ___D___

	16	15	14	13	12	11	10	9	8	7	6	5	4	3	2	1	
Word 100	0	0	0	0	0	0	0	0	0	0	0	0	0	0	0	0	Output
Word 121	1	1	0	0	0	0	0	0	1	1	0	0	0	0	0	1	Step 1
Word 122	0	1	1	0	0	1	0	0	0	0	1	0	1	1	1	1	Step 2
Word 123	0	1	1	0	0	1	0	0	1	0	1	0	1	0	1	1	Step 3
Word 124	1	1	1	0	0	0	1	0	0	1	1	0	1	1	1	0	Step 4

Fig. 12-3

16. Normally, sequencer instructions are retentive. This means that, if you are at step 3 of a 10-step sequencer and power to your system is lost, when power is restored the sequencer will:
a) return to step 3.
b) increment to step 4 automatically.
c) reset to step 1 automatically.
d) advance to step 10 automatically.

16. _____

17. A single sequencer instruction will have an upper limit on:
a) the number of steps that can be programmed.
b) the number of external outputs that can be programmed.
c) the number of times the operating cycle can be actuated.
d) both *a* and *b.*

17. _____

18. When using a time-driven sequencer, the sequencer advances to the next step:
a) when the preset value equals the accumulated value.
b) when the preset value is less than the accumulated value.
c) for every true-to-false transition of the sequencer rung.
d) for every false-to-true transition of the sequencer rung.

18. _____

19. A reset instruction for a sequencer:
a) is never required.
b) is required only if the sequencer is retentive.
c) is required only if the sequencer is nonretentive.
d) is given the same address as the sequencer.

19. _____

20. The sequencer program of Fig. 12-4 is:
a) motor-driven.
b) time-driven.
c) event-driven.
d) gear-driven.

20. _____

21. The sequencer of Fig. 12-4 operates whenever:
a) input A is false.
b) input A is true.
c) the PLC is in the run mode.
d) the PLC is in the program mode.

21. _____

22. When the sequencer program of Fig. 12-4 is operated, the circuit increments automatically through the _____ steps of the sequencer at _____ intervals.
a) five – 3 second
b) six – 3 second
c) five – 3 minute
d) six – 3 minute

22. _____

23. When the sequencer program of Fig. 12-4 is at position 2, which output(s) will be energized?
a) O:2/12
b) O:2/13
c) O:2/12 and O:2/13
d) O:2/14

23. _____

Ladder logic program

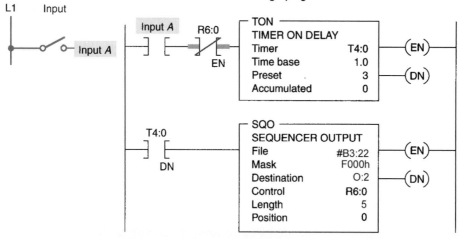

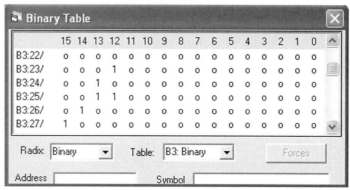

Fig. 12-4

24. A program modification is to be made to the sequencer program of Fig 12-4 that requires outputs O:2/12, O:2/12, O:2/12, and O:2/12 to all be energized at step 3. This would require:
   a) bits 12, 13, 14 and 15 of word B3:22 to be set to 1.
   b) bits 13, 14 and 15 of word B3:23 to be set to 1.
   c) bits 12, 14 and 15 of word B3:24 to be set to 1.
   d) bits 14 and 15 of word B3:25 to be set to 1.

24. _____

25. The sequencer input instruction is true whenever the input data are _____ the data stored in the sequencer file.
   a) equal to
   b) greater than
   c) less than
   d) greater than or equal to

25. _____

26. The sequencer load instruction _____ data from a source to a sequence file.
   a) adds
   b) subtracts
   c) copies
   d) negates

26. _____

27. The length of the bit shift instructions is given in:
a) words.
b) bits.
c) steps.
d) files.

27. _____

28. A common application for a shift register would be:
a) tracking parts.
b) controlling machine or process operations.
c) inventory control.
d) all of these.

28. _____

29. In a word shift register, the data are shifted out _____ bit(s) at a time.
a) 1
b) 2
c) 4
d) 16

29. _____

30. Which instruction is also known as a FIFO (first in/first out)?
a) Synchronous shift register
b) Word to file move
c) File to word move
d) Asynchronous shift register

30. _____

TEST 12•2

Place the answers to the following questions in the answer column at the right.

Answer

1. Mechanical sequencer switches are often referred to as (a) _____ switches, (b) _____ switches, (c) _____ switches, or (d) _____ switches.

1a. _____

1b. _____

1c. _____

1d. _____

2. In the mechanical sequencer switch in Fig. 12-5, contacts interact with the cam so that, in different degrees of rotation of the cam, different contacts (a) _____ and (b) _____.

2a. _____

2b. _____

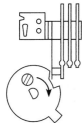

Cam and contact operation

Fig. 12-5

3. In the mechanical sequencer of Question 2, an electric _____ is used to drive the cam.

3. _____

4. Sequencer switches are used whenever a repeatable operating pattern is required. (True or False)

4. _____

5. To program a sequencer, binary information is entered into a series of consecutive memory words referred to as a(n) _____.

5. _____

6. As a programmed sequencer advances through its steps, binary information is transferred from the (a) _____ to the (b) _____.

6a. _____

6b. _____

7. For ease of programming, some PLCs allow the sequencer word file data to be entered using the octal, hexadecimal, or BCD number system. (True or False)

7. _____

8. When a sequencer operates on an entire output word, all outputs associated with the word are required to be controlled by the sequencer. (True or False)

9. Bits of an output word not used by the sequencer can be used elsewhere in your program. (True or False)

10. The _____ word selectively screens out data from the sequencer word file to the output word.

11. Sequencers, like other PLC instructions, are programmed exactly the same for all PLC models. (True or False)

12. The advantage of sequencer programming over the conventional program is the large saving of memory words. (True or False)

13. Sequencer instructions simplify your ladder program. (True or False)

14. The sequencer output instruction sends a predetermined 16-bit word representing the desired output configuration for a specific _____ module.

15. Sequencer instructions are usually nonretentive. (True or False)

16. There is usually no limit to the number of external outputs and steps that can be operated on by a single sequencer instruction. (True or False)

17. Many sequencer instructions reset the sequencer automatically to step 1 on completion of the last sequence step. (True or False)

18. A (n) _____-driven sequencer operates in a manner similar to a mechanical drum switch that increments automatically after a preset time period.

19. The hexadecimal equivalent of the binary number 0011 1111 is _____.

20. With time-driven sequencers, each step functions in a manner similar to timer instructions because it involves a(n) (a) _____ time value and a programmed (b) _____ time value.

8. _____

9. _____

10. _____

11. _____

12. _____

13. _____

14. _____

15. _____

16. _____

17. _____

18. _____

19. _____

20a. _____

20b. _____

21. The _____ parameter is the address of the sequencer file.

21. _____

22. Before a sequencer starts its sequence, we need a starting point where the sequencer is in a neutral position. The start position is all zeroes, representing our neutral position; thus all outputs will be _____.

22. _____

23. The sequencer _____ parameter is where the status bits, length, and instantaneous position are stored.

23. _____

24. Shift registers are often used for _____ parts on a production line.

24. _____

25. Generally, there are two bit shift instructions: bit shift (a) _____ and bit shift (b) _____.

25a. _____

25b. _____

26. With a bit shift register, the status data (1 or 0) is shifted automatically through the register from one bit _____ to the next.

26. _____

27. The sequencer output instruction of Fig. 12-6 will send the data in file (a) _____ out to the (b) _____ output connections. The mask value of 00FF means that only the (c) _____ 8 bits will be sent. The length of the file is specified as 12 (d) _____.

27a. _____

27b. _____

27c. _____

27d. _____

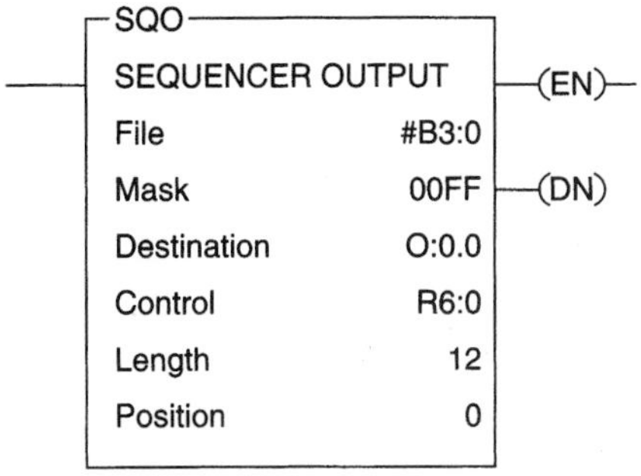

```
┌─SQO─────────────────────
│  SEQUENCER OUTPUT      ├──(EN)─
│  File          #B3:0   │
│  Mask          00FF    ├──(DN)
│  Destination   O:0.0   │
│  Control       R6:0    │
│  Length        12      │
│  Position      0       │
└────────────────────────
```

Fig. 12-6

28. In the bit shift instruction of Fig. 12-7, (a) _____ bits of data starting at (b) _____ will be shifted out the (c) _____ output connection, 1 bit at a time. When all the bits have been shifted out, the Done (DN) bit will be set to d) _____.

```
┌─ BSL ──────────────────┐
│  BIT SHIFT LEFT        │──(EN)—
│  File         #B3:0    │
│  Control      R6:0     │──(DN)
│  Bit Address  O:0.0/4  │
│  Length       128      │
└────────────────────────┘
```

Fig. 12-7

28a. _____

28b. _____

28c. _____

28d. _____

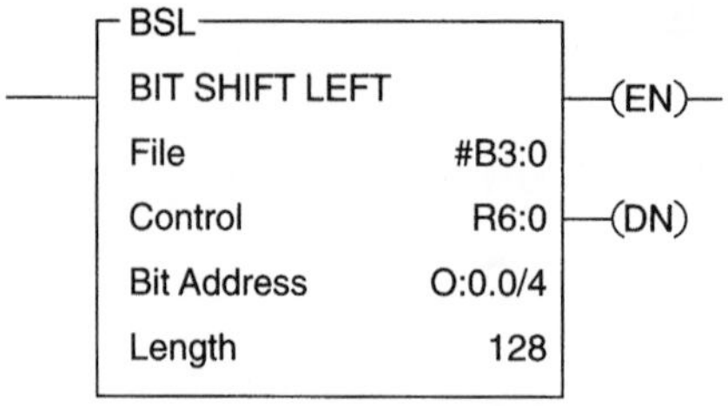

29. State which sequencer instruction (SQO, SQ1, or SQL) would be used if you want to:
   a) Capture reference conditions by manually stepping the machine through its operating sequences and loading I/O storage data into destination files.
   b) Control sequential machine operations by transferring 16-bit data to output image addresses.
   c) Monitor machine operating conditions for diagnostic purposes by comparing 16-bit image data with data in a reference file.

29a. _____

29b. _____

29c. _____

30. Shift registers can be used to control machines or processes where parts are shifted continually from one position to the next. (True or False)

30. _____

31. You can program a shift register instruction to shift 1's or 0's through the register. (True or False)

31. _____

32. If you wanted to produce an external output when a certain bit in a shift register is on, you could program a rung with an EXAMINE IF CLOSED instruction corresponding to the address of the shift register _____ address.

32. _____

33. When you program a shift register instruction, you can shift data left only. (True or False)

33. _____

34. Complete the information for the mask word and sequencer file in Fig. 12-8 so that the sequencer will operate the lamps as shown (dark circle indicates lamp is on).

34. _____

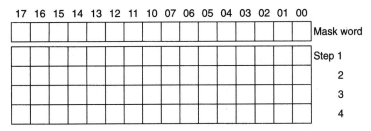

17	16	15	14	13	12	11	10	07	06	05	04	03	02	01	00	
																Mask word
																Step 1
																2
																3
																4

Fig. 12-8

35. For the SQO instruction condition in Fig. 12-9, which outputs will be on?

35. _____

```
┌─SQO────────────────────────┐
│ SEQUENCER OUTPUT          ─(EN)─
│ File            #B10:1
│ Mask            0F0F        ─(DN)─
│ Destination     O:14.0
│ Control         R6:20
│ Length          4
│ Position        2
└────────────────────────────┘
```

External outputs associated with O:14

| 00 |
| 01 |
| 02 |
| 03 |
| 04 |
| 05 |
| 06 |
| 07 |
| 08 |
| 09 |
| 10 |
| 11 |
| 12 |
| 13 |
| 14 |
| 15 |

Sequencer output file #B10:1

Word					Step
B10:1	0000	0000	0000	0000	0
2	1010	0010	1111	0101	1
3	1111	0101	0100	1010	2
4	0101	0101	0101	0101	3
5	0000	1111	0000	1111	4

Fig. 12-9

36. For the BSR instruction in Fig. 12-10, which parameter tells the processor:

a) the instruction's address?
b) the number of bits in the bit array?
c) the source bit address?
d) The location of the bit array?

BSR
BIT SHIFT RIGHT
File
Control
Bit address
Length
(EN)
(DN)

Fig. 12-10

36a. ______________________

36b. ______________________

36c. ______________________

36d. ______________________

37. For the FFL-FFU instruction pair in Fig. 12-11, which parameter tells the processor:

a) the location of the exit word?
b) the location of the next in word?
c) to start at the FIFO file address?
d) the location of the stack?
e) the instruction's address?
f) the maximum number of words you can load?

FFL
FIFO LOAD
Source N60:1
FIFO #N60:3
Control R6:51
Length 64
Position 0
(EN)
(DN)
(EM)
FFU
FIFO UNLOAD
FIFO #N60:3
Destination N60:2
Control R6:51
Length 64
Position 0
(EU)
(DN)
(EM)

Fig. 12-11

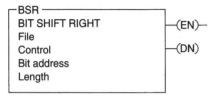

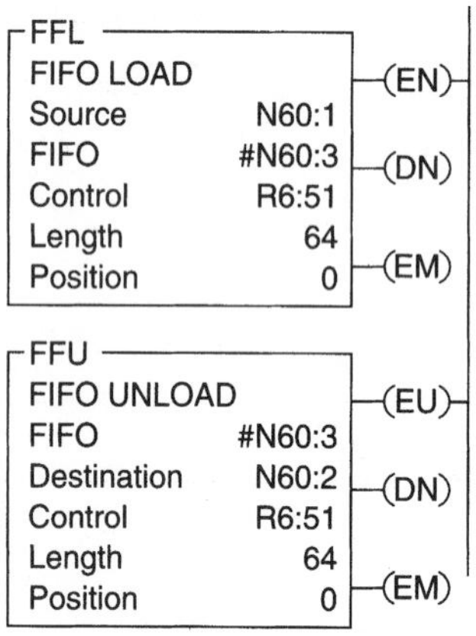

37a. ______________________

37b. ______________________

37c. ______________________

37d. ______________________

37e. ______________________

37f. ______________________

PROGRAMMING ASSIGNMENTS FOR CHAPTER 12

1. For the PLC you will be working with, record each of the following sequencer specifications:
   a) Maximum number that can be programmed
   b) Sequencer addresses
   c) Maximum number of steps that can be programmed per sequencer instruction
   d) Maximum number of external outputs that can be programmed per sequencer instruction

2. Design a simulated sequencer program for the mechanical sequencer in Fig. 12-12, which is used as part of a dishwasher circuit. The data table outlines the sequence of operation of the timer. Each time increment is 45 s. A total of 60 45-s steps is used to complete the 45-min operating cycle. The active circuits column refers to the encircled numbers found on the schematic diagram. Construct the program using any available addresses, switches, and lights on your PLC demonstration panel. After constructing your program on a separate sheet of paper, enter it into the PLC and prove its operation.

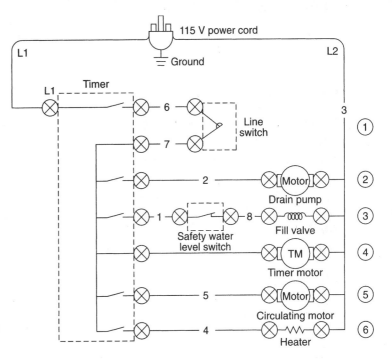

Machine function		Timer increment	Active circuits				
Off		0–1					
First prerinse	Drain	2	1	2	4		
	Fill	3	1		3	4	5
	Rinse	4–5	1			4	5	6
	Drain	6	1	2	4	5	
Prewash	Fill	7	1		3	4	5
	Wash	8–10	1			4	5	6
	Drain	11	1	2	4	5	
Second prerinse	Fill	12	1		3	4	5
	Rinse	13–15	1			4	5	6
	Drain	16	1	2	4		
Wash	Fill	17	1		3	4	
	Wash	18–30	1			4	5	6
	Drain	31	1	2	4	5	
First rinse	Fill	32	1		3	4	5
	Rinse	33–34	1			4	5	6
	Drain	35	1	2	4	5	
Second rinse	Fill	36	1		3	4	5
	Rinse	37–41	1			4	5	6
	Drain	42	1	2	4	5	
Dry	Dry	43–58	1			4		6
	Drain	59	1	2	4		6
	Dry	60	1			4		6

Fig. 12-12

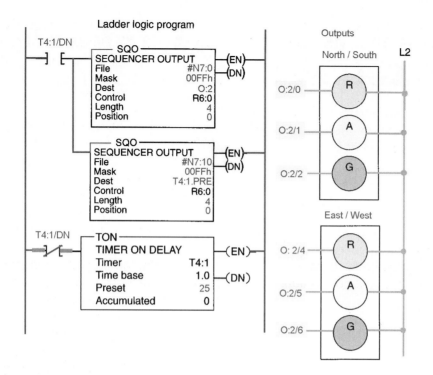

Ladder logic program

Timing Chart

RED (N/S)		GREEN (N/S)	AMBER (N/S)
GREEN (E/W)	AMBER (E/W)	RED (E/W)	

<-------- 25 s ------>X---- 5 s --->X--------- 25 s -------->X--- 5 s -->

Sequencer File #N7:0 Light Cycle Settings

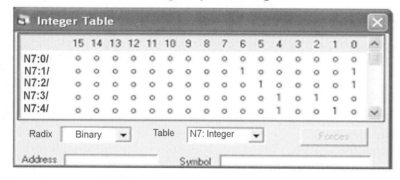

Sequencer File #N7:10 Timer Settings

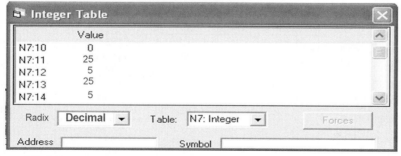

Fig. 12-13

3. Construct a simulated program for the time-driven traffic light sequencer program in Fig. 12-13. This program simulates the operation of two-way traffic lights. After constructing your program on a separate sheet of paper, enter the program into the PLC and prove its operation.

4. Modify the traffic light program of Question 3 to operate as an *event-driven* sequencer. In this application, use a pushbutton to step manually through the different sequencer steps. Enter the modified program into the PLC and prove its operation.

5. Design a PLC sequencer program that provides the following:
   a) The sequencer must step in 5-s intervals
   b) Output No. 1—on all the time the machine is cycling
   c) Output No. 2—on except for steps 3 and 5
   d) Output No. 3—on only in step 3
   e) Output No. 4—on in steps 2 and 4
   f) Output No. 5—on in steps 2, 3, and 4
   g) Output No. 6—on in steps 1 and 5

   Prepare a sequencer instruction data form, I/O connection diagram, and ladder logic program for the circuit. Enter the program into the PLC and prove its operation.

6. Modify the program of Question 5 to provide the following additional control features:
   a) All outputs must stay off and the sequencer must not operate until a start button is pressed.
   b) Once the start button is pressed, the sequencer completes one complete cycle and then stops automatically.
   c) Pushing a stop button resets and stops the sequencer.

Prepare a sequencer instruction data form, I/O connection diagram, and ladder logic program for the circuit. Enter the program into the PLC and prove its operation.

7. Traffic flow on a one-way street is to be controlled by means of a pedestrian pushbutton so that the green traffic light and the Don't Walk pedestrian light are to be normally on at all times when the pedestrian pushbutton is not actuated; and when the pedestrian pushbutton is actuated, the sequencer is started and controls the outputs as follows:
   a) The green traffic light immediately switches off and the amber traffic light switches on to begin to stop the traffic flow—the Don't Walk pedestrian light remains on. Outputs remain in this state for 5 s.
   b) The amber traffic light switches off and the red traffic light switches on—the Don't Walk pedestrian light remains on. Outputs remain in this state for 5 s to ensure that traffic has stopped before pedestrians begin to cross.
   c) The Don't Walk pedestrian light switches off and the Walk pedestrian light switches on—the red traffic light remains on. Outputs remain in this state for 15 s, allowing pedestrians safe passage across the street.
   d) The Walk pedestrian light switches off and the Don't Walk pedestrian light switches on—the red traffic light remains on. Outputs remain in this state for 5 s to ensure that pedestrians are not still crossing the street when the traffic light changes from red to green.

e) The green traffic light switches on and the red traffic light switches off—the Don't Walk pedestrian light remains on. Outputs remain in this state for 30 s to ensure a minimum amount of automobile traffic flow time, even if the walk pushbutton is frequently actuated.

f) The sequencer stops, and the green traffic light and Don't Walk pedestrian light remain on until the pedestrian pushbutton is pressed to start the cycle again.

Prepare a sequencer instruction data form, I/O connection diagram, and ladder logic program that can be used to simulate this traffic control system. Enter the program into the PLC and prove its operation.

8. Design a PLC event-driven sequencer output program for the automatic car wash matrix in Fig. 12-14. Enter the program into the PLC and prove its operation.

Event	Water input	Soap release	Hot wax	Air blower
LS1	1	0	0	0
LS2	1	1	0	0
LS3	1	0	0	0
LS4	0	0	1	0
LS5	0	0	0	1
LS6	0	0	0	0

Fig. 12-14

9. a) Using the sequencer output (SQO) instruction, design a program to turn on outputs according to the 8 steps in Fig. 12-15. Use a switch to step through the table. Enter the program into the PLC and prove its operation.

b) Modify the program to operate continuously by using a recycling timer's done bit to trigger a step in the sequence. Enter the modified program into the PLC and prove its operation.

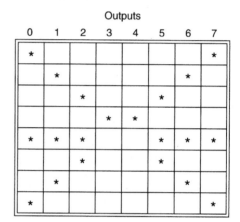

Fig. 12-15

10. Create a file of timer preset values and create a sequencer output file. The file of timers will work in conjunction with the sequencer so that the outputs remain on for a period of time derived from the timer table. The sequencer should remain in a particular step until the timer times out, then proceeds to the next step. Repeat the next preset and step. Enter the program into the PLC and prove its operation.

11. Use a sequencer input (SQI) instruction and a sequencer output (SQO) instruction on the same rung to produce the resulting output according to the table below. Enter the program into the PLC and prove its operation.

Input Data	Output Data
00000001	0000000
00000010	1000000
00000100	0100000
00001000	0010000
00010000	0001000
00100000	0000100
01000000	0000010
10000000	0000001

12. Create an event-driven sequencer output instruction program similar to that described in Fig. 12-14 of the text. Enter the program into the PLC and prove its operation.

13. Create a sequencer input instruction program similar to that described in Fig. 12-16 of the text. Enter the program into the PLC and prove its operation.

14. Create a sequencer load instruction program similar to that described in Fig. 12-20 of the text. Enter the program into the PLC and prove its operation.

15. Create a bit shift left program similar to that described in Fig. 12-24 of the text. Enter the program into the PLC and prove its operation.

16. Create a bit shift right program similar to that described in Fig. 12-25 of the text. Enter the program into the PLC and prove its operation.

17. Create a shift register spray-painting program similar to that described in Fig. 12-26 of the text. Enter the program into the PLC and prove its operation.

18. Create a bit shift operation program for keeping track of carriers flowing through a 16-station machine, similar to that described in Fig. 12-27 of the text. Enter the program into the PLC and prove its operation.

19. Create a FIFO word shift register program similar to that described in Fig. 12-29 of the text. Enter the program into the PLC and prove its operation.

20. Construct a program that will keep track of the presence of parts on a 23-station conveyor line. If a part is placed on the line, then a limit switch connected to input *A* address will close. The conveyor will be indexed by pressing a pushbutton connected to input *B*. An indicator light connected to output *C* will turn on when a part comes off the line. Enter the program into the PLC and prove its operation.

13 PLC INSTALLATION PRACTICES, EDITING, AND TROUBLESHOOTING

TEST 13•1

Choose the letter that best completes the statement.

Answer

1. PLCs are placed within an enclosure to provide protection against:
a) atmospheric conditions.
b) conductive dust.
c) moisture.
d) all of the above.

1. _____

2. For most PLC installations, a NEMA _____ enclosure is recommended.
a) 2
b) 4
c) 8
d) 12

2. _____

3. Typically, PLC systems installed inside an enclosure can withstand a maximum of:
a) 60°C outside the enclosure.
b) 50°C outside the enclosure.
c) 60°C inside the enclosure.
d) 50°C inside the enclosure.

3. _____

4. Malfunctions due to electrical noise interference usually cause:
a) temporary occurrences of operating errors.
b) permanent occurrences of operating errors.
c) temporary loss of memory.
d) permanent loss of memory.

4. _____

5. A good location for a PLC enclosure is close to:
a) the machine or process.
b) large ac motors.
c) high frequency welders.
d) annealing furnaces.

5. _____

6. Proper spacing of components within the enclosure is an important factor in:

    a) determining the amount of input current flow.

    b) determining the amount of output current flow.

    c) lowering the operating temperature within the enclosure.

    d) making use of all available space for mounting of component parts.

6. _____

7. Electrical noise can be coupled into a PLC system:

    a) by an electrostatic field.

    b) through electromagnetic induction.

    c) by a fiberoptic system.

    d) both _a_ and _b_.

7. _____

8. Which of the following would _not_ normally be located within the PLC enclosure?

    a) I/O modules

    b) Limit switch

    c) Master control relay

    d) Isolation transformer

8. _____

9. Under _no_ circumstances should:

    a) fiberoptic and power wiring be run in the same conduit.

    b) fiberoptic and signal wiring be run in the same conduit.

    c) signal wiring and power wiring be run in the same conduit.

    d) all of the above.

9. _____

10. Some input devices may have a small leakage current when they are:

    a) in the on state.

    b) in the off state.

    c) examined for an on condition.

    d) examined for an off condition.

10. _____

11. A leakage problem can occur when connecting an output module to a high-impedance load. This problem can be corrected by connecting:

    a) a bleeder resistor in series with the load.

    b) a bleeder resistor in parallel with the load.

    c) an NO contact in series with the load.

    d) an NC contact in parallel with the load.

11. _____

12. I/O leakage problems usually occur with devices that use:

    a) solid-state switching circuits.

    b) hard contacts.

    c) noise-suppression circuits.

    d) voltage-suppression circuits.

12. _____

13. The authoritative source on grounding requirements for a PLC installation is the:

    a) plant electrician.

    b) plant engineer.

    c) equipment manufacturer.

    d) National Electrical Code.

13. _____

14. In addition to being an important safety measure, proper grounding of a PLC system can:
    a) lower installation costs.
    b) increase the power efficiency of the system.
    c) limit the effects of EMI.
    d) assist in the operation of the MCR.

14. _____

15. Proper grounding procedures for a PLC installation specify that:
    a) all enclosure backplates should be individually grounded to a control ground bus.
    b) paint or nonconductive materials should be scraped away to provide good ground connections.
    c) all ground connections should be made using star washers.
    d) all of the above.

15. _____

16. Which of the following load devices is most likely to require some form of noise or voltage suppression?
    a) Lamp
    b) Heater
    c) Solenoid
    d) LED display

16. _____

17. Excessive line voltage variations to a PLC installation can be corrected by installing a:
    a) constant voltage transformer.
    b) step-down transformer.
    c) step-up transformer.
    d) current transformer.

17. _____

18. A high voltage spike is generated whenever current to a(n):
    a) inductive load is turned off.
    b) inductive load is turned on.
    c) resistive load is turned off.
    d) resistive load is turned on.

18. _____

19. Which of the following function(s) is(are) used when editing a PLC program?
    a) Remove/insert
    b) Point-and-click with a mouse button
    c) Search
    d) All of the above

19. _____

20. In which of the following program modes are modifications executed immediately on entry of the instruction?
    a) Continuous test mode
    b) Single scan test mode
    c) Off-line program mode
    d) On-line program mode

20. _____

21. Which of the following would *not* normally be included as part of a routine preventive maintenance program?
   a) Inspection of I/O field devices
   b) Monitoring the program
   c) Connections to I/O modules
   d) Cleaning of the PLC enclosure

21. _____

22. If an output module fuse blows repeatedly, a probable cause may be:
   a) the module's output current is being exceeded.
   b) the output device may be shorted.
   c) the output field wiring may be shorted.
   d) all of the above.

22. _____

23. Which of the following is *not* normally included in the array of status indicators found on the processor module?
   a) Memory OK
   b) Wiring OK
   c) Battery OK
   d) Power supply OK

23. _____

24. A watchdog timer is used to monitor the:
   a) scan process of the system.
   b) battery voltage.
   c) memory circuits.
   d) dc logic voltage.

24. _____

25-1. The input hardware circuit is suspected as being the source of a PLC problem. If this circuit is *not* at fault, then the status indicator on the input module should be illuminated when the input device is:
   a) on and programmed for an EXAMINE IF CLOSED condition.
   b) on and programmed for an EXAMINE IF OPEN condition.
   c) on, regardless of how it is programmed.
   d) off, regardless of how it is programmed.

25-1. _____

25-2. Assume an open in the *field wiring* is suspected between the output module and output load device. This condition would be confirmed if:
   a) full output voltage was measured at the module and 0 voltage at the load.
   b) full output voltage was measured at the load and 0 voltage at the module.
   c) the output status indicator on the module is on and the load is not operating.
   d) the output status indicator on the module is off and the load is not operating.

25-2. _____

26-1. Which of the choices in Fig. 13-1 indicates a correct status?

26-1. _____

	Input device condition	Input module status indicator	Monitor display status indicator	
			False	True
(a)	Closed — on	Off	—] [—	- ⅄ -
			True	False
(b)	Open — off	Off	-] [-	—⅄—
			False	True
(c)	Open — off	Off	—] [—	- ⅄ -
			False	True
(d)	Closed — on	On	—] [—	- ⅄ -

Fig. 13-1

26-2. Which of the choices in Fig. 13-2 indicates a problem with the wiring to the output device or the output device itself?

26-2. _____

	Output device condition	Output module status indicator	Monitor display status indicator
(a)	De-energized — off	On	True —( )—
(b)	De-energized — off	Off	True —( )—
(c)	Energized — on	On	True —( )—
(d)	De-energized — off	Off	False —( )—

Fig. 13-2

27. Testing between the input terminal and common of a dc input module when the field wiring is short circuited at the field device terminals will cause the dc voltmeter to read:
a) the user power supply voltage.
b) half user power supply voltage.
c) 0.
d) 24-V dc.

27. _____

28. Ground connections should have a resistance value of less than:
    a) 0.1 Ω.
    b) 1.0 Ω.
    c) 10.0 Ω.
    d) 0.5 Ω.

28. _____

29. Most PLC system faults are caused by:
    a) faulty power supplies.
    b) malfunctioning microprocessors.
    c) software failure.
    d) I/O circuitry.

29. _____

30. A single output device has failed while the remainder of the PLC system is functioning normally. The indicator light on the output module indicates that a signal is sent to the output point where the device is connected. You would now:
    a) trace the circuit back through the logic to locate the inputs.
    b) use a programming terminal to call up the rung that controls the output to see if the output coil is on.
    c) check the point where the output device's field wiring is connected to the output rack.
    d) check the input modules for short-circuit conditions.

30. _____

31. The force instructions:
    a) will force data table bits on only.
    b) will force data table bits off only.
    c) if used indiscriminately, could cause haphazard machine operation.
    d) can be used only to force inputs.

31. _____

32. The first step in troubleshooting is to:
    a) identify or describe the faulty operation.
    b) test the process field devices.
    c) test the wiring.
    d) test the I/O modules.

32. _____

33. The I/O module input and output status lights:
    a) are found only on analog modules.
    b) indicate the status of the inputs and outputs.
    c) indicate whether the process field devices are faulty.
    d) are not used in the troubleshooting process.

33. _____

TEST 13•2

Place the answers to the following questions in the answer column at the right.

Answer

1. PLCs are generally placed within an enclosure. (True or False)

1. _____

2. An enclosure is used to shield the controller from electrical (a) _____ and airborne (b) _____.

2a. _____

2b. _____

3. Most PLC installations require additional cooling provisions, not included in the original installation. (True or False)

3. _____

4. PLC malfunctions due to noise usually produce temporary occurrences of operating errors. (True or False)

4. _____

5. Noise may be coupled into PLC power or control lines by a(n) (a) _____ or through a(n) (b) _____.

5a. _____

5b. _____

6. Potential noise generating devices include noninductive resistive loads. (True or False)

6. _____

7. A fiberoptic system is most susceptible to electrical noise. (True or False)

7. _____

8. Running signal wiring and power wiring in the same conduit helps to cut down on electrical noise. (True or False)

8. _____

9. Most solid-state switches will conduct a small amount of leakage current in the _____ state.

9. _____

10. Leakage current can falsely activate a PLC input. (True or False)

10. _____

11. A _____ resistor can be connected to drain off unwanted leakage current.

11. _____

12. Proper grounding of a PLC system is an important safety feature and can also help to limit the effects of _____.

12. _____

13. Where line voltage variations to the PLC are excessive, a(n) _____ transformer can be used to maintain a steady voltage.

13. _____

14. When current in a(n) _____ load is interrupted or turned off, a very high voltage spike is generated.

14. _____

15. A(n) _____ can be connected in reverse-bias across a dc solenoid to suppress voltage spikes.

15. _____

16. Generally, output modules designed to drive inductive loads include suppression networks as part of the module circuit. (True or False)

16. _____

17. Using PLC editing functions, instructions and rungs can be (a) _____ or (b) _____.

17a. _____

17b. _____

18. The _____ function can be used to locate a specified addressed instruction in the processor's memory.

18. _____

19. The mouse button allows you to move through your program from instruction to instruction or from rung to rung. (True or False)

19. _____

20. The on-line programming mode permits the user to change the program during the machine operation. (True or False)

20. _____

21. When in the continuous test mode, the processor operates the user program without _____ any outputs.

21. _____

22. In the _____ test mode, the processor makes one scan of the user program each time the instruction is activated.

22. _____

23. All field I/O devices should be inspected periodically to ensure that they are properly adjusted as an important part of a PLC's _____ maintenance program.

23. _____

24. The first step in the troubleshooting of a PLC system is to identify the (a) _____ and its (b) _____.

24a. _____

24b. _____

25. One of the diagnostic checks carried out by the processor is the proper operation of all I/O devices. (True or False)

25. _____

26. The watchdog timer is a separate timing circuit that must be set and reset by the _____.

26. _____

27. If a processor hardwire malfunction occurs, the watchdog timer circuit will _____ and halt the operation of the PLC.

27. _____

28. Usually, each I/O device has at least two status indicators. One of these indicators is on the I/O (a) _____, the other is provided by the programming device (b) _____.

28a. _____

28b. _____

29. The status indicator on an input module will normally be illuminated if the input device is off and examined for an off condition. (True or False)

29. _____

30. The programming device monitor normally indicates a true instruction if the addressed input device is off and examined for an off condition. (True or False)

30. _____

31. The shield of PLC field wiring is normally grounded at both ends to prevent ground-loop interference. (True or False)

31. _____

32. Lack of surge suppression on inductive loads may contribute to processor faults. (True or False)

32. _____

33. The suppression device is wired in _____ with the load device.

33. _____

34. Surge suppression is also known as _____.

34. _____

35. Data _____ is a feature that allows you to display data from anywhere in the data table.

35. _____

36. The contact histogram function allows you to view the _____ history of a data table value.

36. _____

37. Most companies use (a) _____ and (b) _____ protection to ensure that equipment does not operate while maintenance and repairs are conducted.

37a. _____

37b. _____

38. The temporary end instruction is used when you want to change the amount of logic being scanned. (True or False)

38. _____

39. The suspend instruction is used to trap and identify specific conditions. (True or False)

39. _____

40. It is rare for sensors and actuators connected to the I/O of the process to fail. (True or False)

40. _____

PROGRAMMING ASSIGNMENTS FOR CHAPTER 13

1. Refer to the user's manual for the PLC you are using and outline the different self-detection diagnostic features of the processor.

2-1. Measure and record the input leakage current from a solid-state input device.

2-2. Measure and record the output leakage current from a solid-state output module.

3-1. Connect a relay coil or solenoid in series with a dc source and pushbutton. With an oscilloscope connected across the coil, observe the voltage spikes produced when the pushbutton is operated on and off.

3-2. Properly connect a diode across the coil to suppress the voltage spikes. With an oscilloscope connected across the coil, observe how the voltage spikes have been reduced. On a separate sheet of paper, draw the shape of the waveform seen before and after the diode was connected.

4-1. Construct a simulated program for the program editing and data control exercise in Fig. 13-3 using any available addresses, switches, and lights on your PLC demonstration panel. Have the program checked by your instructor after each editing change. Enter the original program into the PLC and prove its operation.

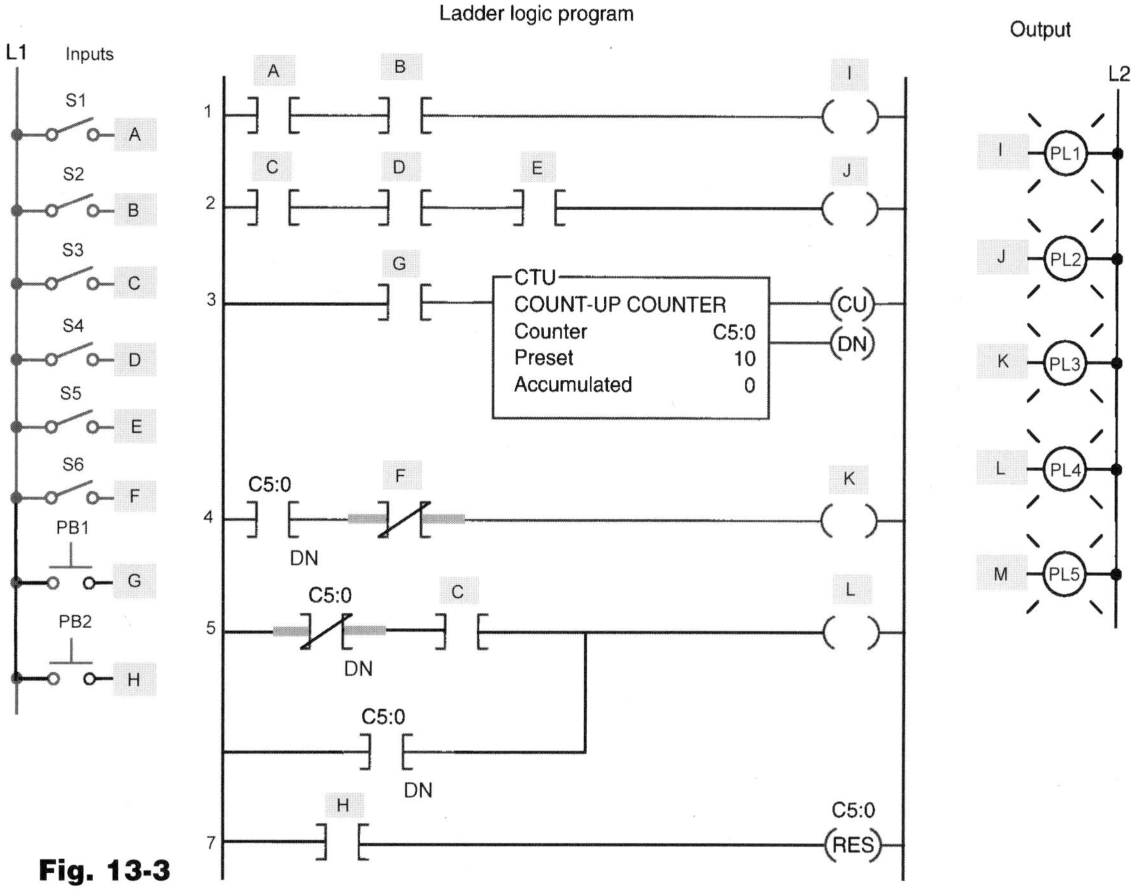

Fig. 13-3

4-2. Enter an additional rung following rung No. 4. This new rung is to examine the status bit of output *I* for an off condition and energize output *M* when the rung condition is true.

4-3. Remove the instruction from rung No. 2 that examines address E for an EXAMINE IF CLOSED condition.

4-4. Place the EXAMINE IF CLOSED instruction (E) that was removed in Question 4-3 into rung No. 6. Insert this EXAMINE IF CLOSED instruction in parallel with the existing EXAMINE IF CLOSED instruction H.

4-5. While in the run mode, change the preset value of the counter from 10 to 25.

4-6. Remove rung No. 5 of the original program.

4-7. While in the run mode, force output *K* to an on condition.

4-8. While in the run mode, force output *J* to an off condition.

5. Enter the original program of Question 4 into the PLC and complete the troubleshooting exercises that follow. Have your instructor simulate each type of fault. Demonstrate your ability to identify the source of each problem systematically.
   a) Defective input module
   b) Defective output module
   c) Blown fuse in an output module
   d) Shorted input device
   e) Open input device
   f) Open output device
   g) Open in field wiring to an input device
   h) Open in field wiring to an output device

PROCESS CONTROL AND DATA ACQUISITION SYSTEMS

TEST 14•1

Choose the letter that best completes the statement.   Answer

1. A continuous process is: **1.** _____
   a) one that never shuts down.
   b) used only for simple tasks.
   c) one in which raw materials enter the process and an identifiable
      product exits the process.
   d) used only with computers.

2. Assume two ingredients are added together, processed, and then   **2.** _____
   stored. This would be an example of a(n):
   a) batch process.
   b) continuous process.
   c) individual product-producing process.
   d) discrete product-producing process.

3. A distributive control system (DCS):   **3.** _____
   a) permits the distribution of the processing task among several
      loop controllers.
   b) always utilizes a single large computer.
   c) will stop the whole process if one control element fails.
   d) is the least flexible type of control system.

4. Components of a control system may include: **4.** _____
   a) sensors.
   b) actuators.
   c) amplifiers.
   d) all of the above.

5. Signal conditioning involves: **5.** _____
   a) converting electrical signals into physical action.
   b) inputs from a human to set up the starting conditions.
   c) converting input and output signals to a usable form.
   d) all of the above.

6. Which of the following devices could be classified as a sensor?   6. _____
   a) Thermistor
   b) Relay
   c) Clutch
   d) All of the above

7. Which of the following devices could be classified as an actuator?   7. _____
   a) Motor
   b) Heater
   c) Fan
   d) All of these

8. Compared to an open-loop system, a closed-loop system in general is: 8. _____
   a) more accurate.
   b) more complex.
   c) more expensive.
   d) all of these.

9. Types of process control can be characterized as: 9. _____
   a) open and closed loop.
   b) closed and voltage loop.
   c) open and current loop.
   d) set point and current loop.

10. The set point refers to: 10. _____
   a) input that determines the operating point for the process.
   b) a process variable that is monitored continually.
   c) a process error that is uncontrolled.
   d) all of these.

11. Compared to open-loop control, closed-loop control: 11. _____
   a) is more costly.
   b) is more difficult to calibrate.
   c) requires feedback concerning the status of the process.
   d) all of these.

12. Stepper-motor control is an example of: 12. _____
   a) closed-loop control.
   b) open-loop control.
   c) variable speed control.
   d) variable torque control.

13. The error signal in a closed-loop control system is: 13. _____
   a) always a positive value.
   b) always a negative value.
   c) the difference between the set point and feedback signal.
   d) the sum of the set point and feedback signal.

14. Which of these controller types provides the fastest response to a system error?
   a) PID
   b) On/off
   c) Proportional plus integral
   d) Proportional plus derivative

14. _____

15. Time proportioning control refers to:
   a) linear movement of the final control element.
   b) varying the ratio of on time to off of the final control element.
   c) the integral action of a controller.
   d) the derivative action of a controller.

15. _____

16. Proportional control:
   a) is used to send a full-on or full-off signal to the controller.
   b) creates a permanent residual error in the operating point of a controlled variable.
   c) is used with sudden signal changes that can drastically affect system operation.
   d) is used when offset signals must be taken into account.

16. _____

17. The integral action responds to:
   a) the size and time duration of the error signal.
   b) the speed at which the error signal is changing.
   c) proportional bandwidth.
   d) proportional gain.

17. _____

18. The derivative action responds to:
   a) the size and time duration of the error signal.
   b) the speed at which the error signal is changing.
   c) proportional bandwidth.
   d) proportional gain.

18. _____

19. A PID controller:
   a) is tuned using a signal generator.
   b) is factory-tuned for optimum performance.
   c) must be custom-tuned to each process.
   d) both *a* and *b.*

19. _____

20. A typical signal conditioning requirement for a thermistor is:
   a) cold junction compensation.
   b) current or voltage excitation.
   c) rectification.
   d) attenuation.

20. _____

21. Analog or signal grounds should:
   a) never be connected to each other.
   b) never be connected to power ground.
   c) also be connected to power ground.
   d) both *a* and *b.*

21. _____

22. The save interval for data logging applications:
    a) is constant.
    b) varies with time.
    c) varies as a function of certain measurements.
    d) all of these.

22. _____

23. A(n) _____ allows a signal to pass from its source to the measurement device without a physical connection.
    a) filter
    b) isolator
    c) graphic user interface
    d) shielded cable

23. _____

24. _____ can be thought of as how closely a quantity can be measured.
    a) Resolution
    b) Real time
    c) Sample-and-hold
    d) Track-and-hold

24. _____

25. A(n)_____ accepts several signals at once and allows the user to pick the signal to be examined.
    a) amplifier
    b) rectifier
    c) multiplexer
    d) converter

25. _____

TEST 14•2

Place the answers to the following questions in the answer column at the right.

Answer

1. A continuous process involves the flow of product material from one section of the process to another. (True or False)

1. _____

2. In a(n) _____ process, a set amount of product is received and then some operation is performed on the product.

2. _____

3. With most automated machines, the operator merely sets up operation and initiates a start. (True or False)

3. _____

4. _____ control is used when several machines are controlled by one controller.

4. _____

5. _____ control involves two or more computers communicating with each other to accomplish the complete control task.

5. _____

6. One disadvantage of centralized control is that, if the main controller fails, the whole process is stopped. (True or False)

6. _____

7. Distributive control systems (DCS) use one controller for all the processing tasks. (True or False)

7. _____

8. Process control can be defined as the functions and operations necessary to change a material either (a) _____ or (b) _____.

8a. _____

8b. _____

9. Actuators convert physical information into electrical signals. (True or False)

9. _____

10. A pushbutton switch could be classified as a type of operator-machine interface. (True or False)

10. _____

11. Signal _____ involves converting input and output signals to a usable form.

11. _____

12. A sensor could be classified as a type of controller. (True or False)

12. _____

13. Control systems can be classified as
(a) _____-loop or (b) _____-loop.

13a. _____

13b. _____

14. A(n) _____-loop system is one in which the output of a process affects the input control signal.

14. _____

15. Sensors convert physical information into _____ signals.

15. _____

16. In an open-loop control system, the controller receives no information concerning the status of the process. (True or False)

16. _____

17. Closed-loop control contains a feedback element. (True or False)

17. _____

18. On/off control eliminates hunting. (True or False)

18. _____

19. With on/off control, the measured variable will _____ around the set point.

19. _____

20. _____ proportioning varies the ratio of on time to off.

20. _____

21. Proportioning action occurs within a proportional _____ around the set-point temperature.

21. _____

22. The operation of a proportional controller leads to process deviation known as _____.

22. _____

23. Integral action eliminates steady-state error. (True or False)

23. _____

24. Derivative action responds to the _____ at which the error signal is changing.

24. _____

25. A PID controller produces an output that depends on the (a) _____, (b) _____, (c) _____ of the system error signal.

25a. _____

25b. _____

25c. _____

26. The _____ input determines the desired operating point for a process.

26. _____

27. With on/off control, the controller will never keep the final control element in a(n) _____ position.

27. _____

28. A PID controller must be custom-tuned to each process being controlled. (True or False)

28. _____

29. Fuzzy logic uses artificial intelligence to continuously readjust PID tuning parameters. (True or False)

29. _____

30. With reference to the block diagram described in the text, identify the indicated parts of the open-loop control system shown in Fig. 14-1.

30a. _____

30b. _____

30c. _____

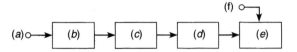

30d. _____

30e. _____

Fig. 14-1

30f. _____

31. Identify the lettered signals of the closed-loop control system shown in Fig. 14-2.

31a. _____

31b. _____

31c. _____

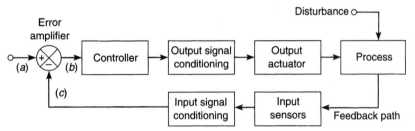

Fig. 14-2

32. With reference to the block diagram described in the text, identify the lettered parts of the PID control system block diagram shown in Fig. 14-3.

32a. _____

32b. _____

32c. _____

32d. _____

32e. _____

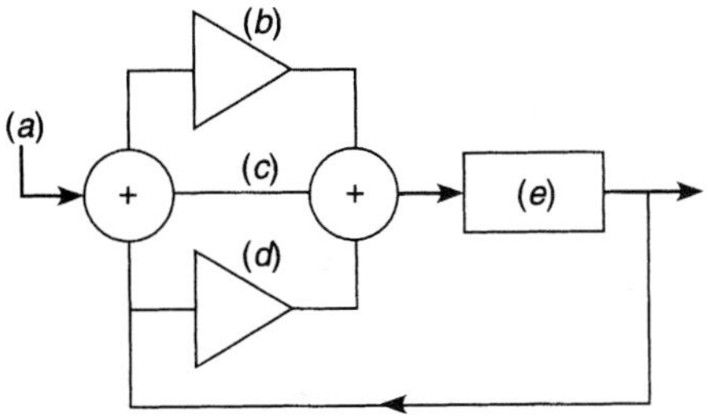

Fig. 14-3

33. With reference to the block diagram described in the text, identify each lettered block of the data acquisition and control system block diagram shown in Fig. 14-4.

33a. _____

33b. _____

33c. _____

33d. _____

33e. _____

33f. _____

33g. _____

33h. _____

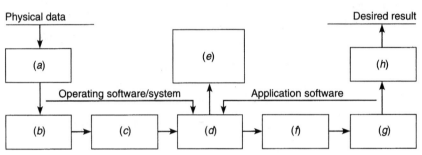

Fig. 14-4

34. Data acquisition is the (a) _____, (b) _____, and (c) _____ of information by a computer-based system.

34a. _____

34b. _____

34c. _____

35. Data acquisition systems can usually be programmed to save various values, which include (a) _____, (b) _____, (c) _____, and (d) _____.

35a. _____

35b. _____

35c. _____

35d. _____

36. Ground loops can produce inaccurate data. (True or False)

36. _____

37. The accuracy and resolution of a measurement system are equal. (True or False)

37. _____

38. A variable save rate could be set up to save data only when something changes. (True or False)

38. _____

For questions 39 to 55, match each term with its definition. Place the letter from the definitions list in the answer column.

DEFINITIONS

a. A false, lower-frequency component that appears in sampled data acquired at too low a sampling rate.
b. Refers to how often the computer checks the measured value.
c. A signal that is not connected to any other systems or voltages.
d. A signal that does not require a reference quantity.
e. A signal that is continuous and has an infinite number of possible values.
f. A property of an operating system in which several processes can be run simultaneously.
g. Takes a very fast "snapshot" of the input voltage and freezes it.
h. Is used to start and/or stop data acquisition.
i. The transfer of data to and/or from a computer system.
j. Data that is processed as it is acquired instead of being accumulated and processed later.
k. A device that accepts several signals at once and allows the user to pick the signal to be examined.
l. A device used to increase the magnitude of a signal.
m. Changes a signal to make it easier to measure or more stable.
n. A signal that has a specific number of possible values.
o. A device used to convert a mechanical or electrical signal into a specific signal such as a 4- to 20-mA loop or a 0- to 5-V signal.
p. The factor by which a signal is amplified.
q. Refers to the process of measuring a signal without direct electrical connection.

TERMS

39. Analog   **39.** _____

40. Transducer   **40.** _____

41. Input/output   **41.** _____

42. Alias   **42.** _____

43. Gain   **43.** _____

44. Absolute signal   **44.** _____

45. Isolation   **45.** _____

46. Multiplexer   **46.** _____

47. Sample and hold   **47.** _____

48. Real time   **48.** _____

49. Control update interval   **49.** _____

50. Floating signal   **50.** _____

51. Digital

52. Multitasking

53. Amplifier

54. Triggering

55. Signal conditioner

51. _____

52. _____

53. _____

54. _____

55. _____

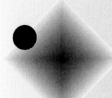

PROGRAMMING ASSIGNMENT
FOR CHAPTER 14

1. Using the internet and whatever manufacturers' catalogues are available to you, report on the specifications and features of a programmable controller PID loop module. Your report should include the following:
   - Description of its operation
   - Special features and options
   - All specifications

15 COMPUTER-CONTROLLED MACHINES AND PROCESSES

TEST 15•1

Choose the letter that best completes the statement.   Answer

1. Disk drives are: 1. _____
   a) used as input devices.
   b) used as output devices.
   c) devices that can only store information.
   d) both *a* and *b.*

2. Disk drives are: 2. _____
   a) labeled using letters of the alphabet.
   b) called either the primary or secondary drives.
   c) given names such as left and right, or top and bottom.
   d) labeled according to the manufacturer.

3. It is common practice to follow the disk drive identification letter by: 3. _____
   a) a semicolon (;).
   b) a period (.).
   c) a colon (:).
   d) a space.

4. To use application software: 4. _____
   a) you must be a programmer.
   b) system software must be loaded into your computer.
   c) no system software can be in your computer.
   d) you need to program it first.

5. OS stands for: 5. _____
   a) operating system.
   b) options setting.
   c) output setting.
   d) none of the above.

6. The purpose of having folders or directories is to: 6. _____
   a) help keep files sorted on the disk.
   b) keep files of similar topics together.
   c) organize disk files.
   d) all of these.

7. The information on floppy disks is stored in:
   a) heads and slots.
   b) sectors and slots.
   c) slots and tracks.
   d) tracks and sectors.

7. _____

8. The information stored on a floppy disk can be corrupted by:
   a) touching the disk surface.
   b) exposing the disk to magnetic fields.
   c) bending the disk.
   d) all of these.

8. _____

9. The four controller areas of computer-integrated manufacturing are:
   a) plant, area, cell, and device.
   b) LAN, MAP, CIM, and DOS.
   c) P, PI, PD, and PID.
   d) horizontal, vertical, circular, and spherical.

9. _____

10. A _____ cell can be defined as a group of equipment integrated to perform a unit of the manufacturing process.
   a) related
   b) lab
   c) work
   d) computer

10. _____

11. For computer-integrated manufacturing, all the different devices must be tied together using a common:
   a) bus.
   b) wire.
   c) protocol.
   d) central control device.

11. _____

12. Open communications networks:
   a) are based on standards developed through industry associations.
   b) do not require that you to buy all componets from a single supplier.
   c) do not use a proprietary protocol.
   d) all of these.

12. _____

13. Bridges are used to:
   a) translate from one network-access scheme to another.
   b) perform data transmission control.
   c) bridge the gap between inout and output devices.
   d) all of these.

13. _____

14. Which of the following is NOT an open network?
   a) DeviceNet
   b) Ethernet
   c) DH-485
   d) Fieldbus

14. _____

15. Feedback is used in numerical control systems to:
   a) improve system stability.
   b) ensure that the expected machine position is the same as the actual position.
   c) reduce the system cost.
   d) increase the number of different types of operations that can be performed.

15. _____

16. Numerical control:
   a) is a method for controlling the operation of a machine by a set of instructions.
   b) applies only to milling machines.
   c) is a method of counting the number of parts produced.
   d) all of these.

16. _____

17. A number of holes are to be drilled in a workpiece using a CNC machine. The type of programming used would be:
   a) point-to-point.
   b) continuous path.
   c) ladder logic.
   d) relay logic.

17. _____

18. The development of a numerical control program requires defining the workpiece within a given:
   a) work envelope.
   b) schematic diagram.
   c) coordinate system.
   d) machine radius.

18. _____

19. The term *nonservo robots* refers to:
   a) hydraulic actuators.
   b) pneumatic actuators.
   c) open-loop systems.
   d) closed-loop systems.

19. _____

20. Bang-bang robots are set up for a task by:
   a) grasping the manipulator and moving it manually from position to position.
   b) adjusting mechanical limit stops.
   c) using high-level computer languages.
   d) all of these.

20. _____

21. Using the robot teach pendant to move a manipulator from point to point is known as:
   a) teach-through programming.
   b) walk-through programming.
   c) interpolation.
   d) protocol.

21. _____

22. The fundamental job of a LAN is to provide _____ between devices.
    a) communication
    b) connections
    c) isolation
    d) protection

22. _____

23. The rate at which a character can be transmitted along a communications line is called:
    a) LAN.
    b) node.
    c) baud.
    d) protocol.

23. _____

24. In the _____ method of data transmission, 1 bit is transferred at a time.
    a) open-loop
    b) closed-loop
    c) serial
    d) parallel

24. _____

25. A _____ bit is used for error detecting.
    a) parity
    b) digital
    c) negative
    d) positive

25. _____

26. Which of the following is used for data transmission between two distant points?
    a) Modem
    b) Telephone
    c) Multiplexer
    d) Demultiplexer

26. _____

27. _____-duplex transmission allows the transmission of data in both directions simultaneously.
    a) Full
    b) Half
    c) Analog
    d) Digital

27. _____

TEST 15•2

Place the answers to the following questions in the answer column at the right.

Answer

1. A computer is made up of (a) _____ and (b) _____ components.

   1a. _____

   1b. _____

2. Floppy drives allow information to be stored and read from nonremovable hard disks. (True or False)

   2. _____

3. A floppy disk must be _____ before any information can be stored on it.

   3. _____

4. Exposure of a computer disk to electromagnetic fields can corrupt the information stored on the disk. (True or False)

   4. _____

5. The hard drive of a computer is normally referred to as the _____ drive.

   5. _____

6. A hard disk drive uses hard disks, which the user can insert and remove easily to produce new programs for the computer. (True or False)

   6. _____

7. Hardware is the tangible part of your computer. (True or False)

   7. _____

8. Software consists of flexible connecting cables and printed output. (True or False)

   8. _____

9. The easiest way to get the operating system into your computer is by programming it through the keyboard. (True or False)

   9. _____

10. The process of loading the operating system into your computer is called booting. (True or False)

    10. _____

11. Windows is a(n) _____-based interface.

    11. _____

12. The serial port is most often used for a printer connection. (True or False)

    12. _____

13. Match the following computer hardware components with their most current description.

COMPONENTS:
Microprocessor chip
Motherboard
Expansion slots
RAM chips
Floppy drives
Power supply
Peripheral cards
ROM chips
Hard drive

DESCRIPTION:

a) Memory programmed at the factory that cannot be changed.

b) Allows information to be stored and read from removable disks.

c) Converts the ac line voltage to the dc voltages needed by the computer system.

d) Interconnects the computer to input and output devices.

e) Used for interconnecting other circuits to the motherboard.

f) Allows information to be stored and read from nonremovable disks.

g) Interprets the instructions for the computer.

h) Memory used to store computer programs and interact with them.

i) Holds and electrically interconnects the major sections of the entire computer system.

13a. _____

13b. _____

13c. _____

13d. _____

13e. _____

13f. _____

13g. _____

13h. _____

13i. _____

14. In general, identify each type of port shown in Fig. 15-1.

14a. _____

14b. _____

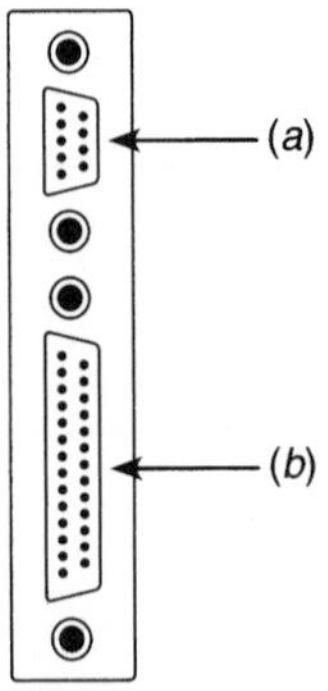

— (a)

— (b)

Fig. 15-1

15. A personal computer can be connected to monitor and control several different manufacturing operations. (True or False)

15. _____

16. A work cell could include CNC, robotic, and PLC control components. (True or False)

16. _____

17. Computer-integrated manufacturing systems provide individual machines with (a) _____ and (b) _____ so they can be integrated into a single system.

17a. _____

17b. _____

18. All manufacturing machines use the same signal protocol. (True or False)

18. _____

19. Types of local area network topologies include (a) _____, (b) _____, and (c) _____.

19a. _____

19b. _____

19c. _____

20. In a master-slave network communication system, direct communication between slave devices is not possible. (True or False)

20. _____

21. A peer-to-peer network system is often described as a(n) _____ passing system.

21. _____

22. Identify each lettered network shown in Fig. 15-2.

22a. _____

22b. _____

22c. _____

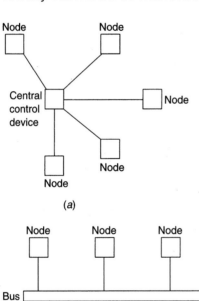

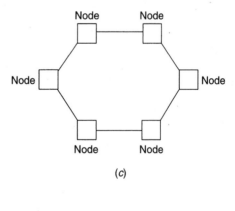

Fig. 15-2

23. Point-to-point programming involves _____ movements.

23. _____

24. The electronic accuracy of a numerical control system is better than the mechanical accuracy. (True or False)

24. _____

25. A set of _____ make up the NC program.

25. _____

26. Today, most NC programs are stored on punched paper tape. (True or False)

26. _____

27. If an NC machine is the three-axis type, the location address of the tool is prefixed with the letter (a) _____, (b) _____, or (c) _____.

27a. _____

27b. _____

27c. _____

28. The programming of NC machines can be categorized as (a) _____ or (b) _____.

28a. _____

28b. _____

29. CNC machines use _____-based control equipment.

29. _____

30. CNC programming consists of taking information from a(n) _____ and converting this information into a computer program.

30. _____

31. Robots with cylindrically shaped work envelopes are always six-axis devices. (True or False)

31. _____

32. AC servo motors are easier to maintain than are dc servo motors. (True or False)

32. _____

33. A robot can be programmed by means of its teach pendant. (True or False)

33. _____

34. A manipular is another name for a robot _____.

34. _____

35. The axis of a robot may be (a) _____, (b) _____, or (c) _____.

35a. _____

35b. _____

35c. _____

36. Using the list provided below, identify each lettered axis of the robot arm shown in Fig. 15-3.

Yaw
Pitch
Shoulder
Roll
Swivel
Arm sweep

36a. _____

36b. _____

36c. _____

36d. _____

36e. _____

36f. _____

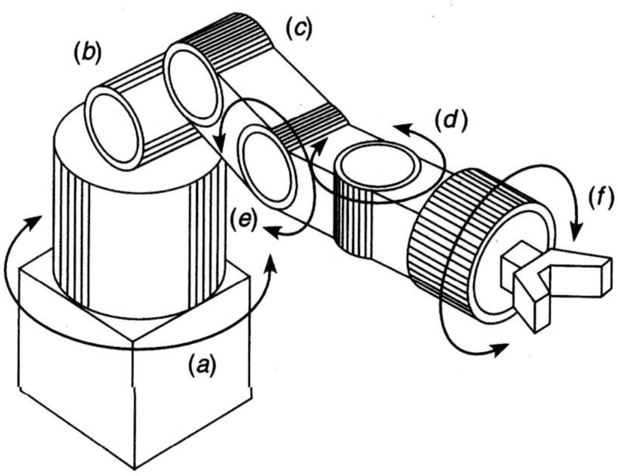

Fig. 15-3

37. The basic industrial robot is an arm that moves to perform operations. (True or False)

37. _____

38. The robot is a series of mechanical links driven by _____.

38. _____

39. The reach of the robot can be defined as the _____.

39. _____

40. The end effector of a robot is always some type of gripper. (True or False)

40. _____

41. The power required to operate the manipulator of a robot can be (a) _____, (b) _____, or (c) _____.

41a. _____

41b. _____

41c. _____

42. Robot manufacturers normally have their own specialized program language for use with their robots. (True or False)

42. _____

43. Three different types of optical disks are (a) _____, (b) _____, and (c) _____.

43a. _____

43b. _____

43c. _____

44. The terms directory and folder are completely interchangeable. (True or False)

44. _____

45. The data transfer rate of a USB port is faster than that of serial or parallel port. (True or False)

45. _____

46. Open network systems limit what devices can be directly connected to the system to that of the manufacturer. (True or False)

46. _____

47. EtherNet/IP is a proprietary industrial network standard. (True or False)

47. _____

PROGRAMMING ASSIGNMENTS FOR CHAPTER 15

1. Network two PLCs using whatever PLC network system is available to you.

2. Using whatever instructional CNC machine is available to you, take information from a part drawing and develop a computer program that will produce the part. Have your program checked by the instructor. Load your program into the CNC controller and operate the machine to produce the part.

3. Program the instructional robot to perform a specialized task defined by the instructor. Program the task using a teach pendant. Repeat the process using a computer and whatever program language the robot understands. Have each program evaluated by the instructor.